THE ACCESSORY DIGESTIVE ORGANS

ATLAS *of* TUMOR RADIOLOGY

PHILIP J. HODES, M.D., *Editor-in-Chief*

Sponsored by

THE AMERICAN COLLEGE OF RADIOLOGY

—*with the cooperation of:*
AMERICAN CANCER SOCIETY
AMERICAN ROENTGEN RAY SOCIETY
CANCER CONTROL PROGRAM, USPHS
EASTMAN KODAK COMPANY
JAMES PICKER FOUNDATION
RADIOLOGICAL SOCIETY OF NORTH AMERICA

THE ACCESSORY DIGESTIVE ORGANS

by

ROBERT E. WISE, M.D.

Chairman, Department of Diagnostic Radiology, Lahey Clinic Foundation
Chairman, Department of Radiology, New England Baptist Hospital
Clinical Professor of Radiology, Boston University School of Medicine

AUSTIN P. O'KEEFFE, M.B., B.Ch.

Radiologist, Exeter Hospital, Exeter, New Hampshire
Teaching Associate, Department of Diagnostic Radiology, Lahey Clinic Foundation

YEAR BOOK MEDICAL PUBLISHERS · INC.
35 EAST WACKER DRIVE · CHICAGO

COPYRIGHT © 1975 BY THE AMERICAN COLLEGE OF RADIOLOGY
20 North Wacker Drive, Chicago, Illinois 60606

Library of Congress Catalog Card Number: 74-18609
International Standard Book Number: 0-8151-9333-5

PRINTED IN U.S.A.

Editor's Preface

In 1960, the Committee on Radiology of the National Research Council began to consider the preparation of a tumor atlas for radiology similar in concept to the Armed Forces Institute of Pathology's "Atlas of Tumor Pathology." So successfully had the latter filled a need in pathology that it seemed reasonable to establish a similar resource for radiology. Therefore a subcommittee of the Committee on Radiology was appointed to study the concept and make recommendations.

That original committee, made up of Dr. Russell H. Morgan (Chairman), Dr. Marshall H. Brucer and Dr. Eugene P. Pendergrass, reported that a need did indeed exist and recommended that something be done about it. That report was unanimously accepted by the parent committee.

Soon thereafter, there occurred a normal change of the membership of the Committee on Radiology of the Council. This was followed by a change of the "Atlas" subcommittee, which now included Dr. E. Richard King (Chairman), Dr. Leo G. Rigler and Dr. Barton R. Young. To this new subcommittee was assigned the task of finding how the "Atlas" was to be published. Numerous avenues were explored; none seemed wholly satisfactory.

With the passing of time, it became increasingly apparent that the American College of Radiology had to be brought into the picture. It had prime teaching responsibilities; it had a Commission on Education; it seemed the logical responsible agent to launch the "Atlas." Confident of the merits of this approach, the entire Committee on Radiology of the Council became involved in focusing the attention of the American College of Radiology upon the matter.* In 1964, as the result of their persuasiveness, the Board of Chancellors of the American College of Radiology named an ad hoc committee to explore and define the scholarly scope of the "Atlas" and the probable costs. In 1965, the ad hoc committee recommended that the College sponsor and publish the "Atlas." Accordingly, an Editorial Advisory Committee was chosen to work within the Commission on Education with authority to select an Editor-in-Chief. At the same time, the College provided funds for starting the project and began representations for grants-in-aid without which the "Atlas" would never be published.

No history of the "Atlas of Tumor Radiology" would be complete without specific recording of the generous response of the several radiological

---
* At that time, the Committee on Radiology included, in addition to the subcommittee, Drs. John A. Campbell, James B. Dealy, Jr., Melvin M. Figley, Hymer L. Friedell, Howard B. Latourette, Alexander Margulis, Ernest A. Mendelsohn, Charles M. Nice, Jr., and Edward W. Webster.

societies, as well as the private and Federal granting institutions whose names appear on the title page and below among our acknowledgments. It was their tangible evidence of confidence in the project that provided everyone with enthusiasm and eagerness to achieve our goal.

The "Atlas of Tumor Radiology" includes all major organ systems. It is intended to be a systematic body of pictorial and written information dealing with the roentgen manifestations of tumors. No attempt has been made to provide an atlas equivalent of a medical encyclopedia. Nevertheless the "Atlas" is designed to serve as an important reference source and teaching file for all physicians, not radiologists alone.

The fourteen volumes of the "Atlas" are: *The Hemopoietic and Lymphatic Systems,* by Gerald D. Dodd and Sidney Wallace; *The Bones and Joints,* by Gwilym S. Lodwick (published); *The Chest,* by Roy R. Greening and J. Haynes Heslep (published); The Gastrointestinal Tract: *The Esophagus and Stomach* (published) and *The Duodenum, Small Intestine and Colon* (published), by George N. Stein and Arthur K. Finkelstein; *The Kidney,* by John A. Evans and Morton A. Bosniak (published); *The Lower Urinary Tract, Adrenals and Retroperitoneum,* by Morton A. Bosniak, Stanley S. Siegelman and John A. Evans; *The Breast,* by David M. Witten (published); *The Head and Neck,* by Gilbert H. Fletcher and Bao-Shan Jing (published); *The Brain and Eye,* by Ernest W. Wood, Juan M. Taveras and Michael S. Tenner; *The Female Reproductive System,* by G. Melvin Stevens (published); *The Endocrines,* by Howard L. Steinbach and Hideyo Minagi (published); *The Accessory Digestive Organs,* by Robert E. Wise and Austin P. O'Keeffe (published); and *The Vertebral Column,* by Bernard S. Epstein (published).

Some overlapping of material in several volumes is inevitable, for example, tumors of the female generative system, tumors of the endocrine glands and tumors of the urinary tract. This is considered to be an asset. It assures the specialist completeness in the volume or volumes that concern him and provides added breadth and depth of knowledge for those interested in the entire series.

The broad scope of the "Atlas of Tumor Radiology" has precluded its preparation by a single or even several authors. To maintain uniformity of format, rather rigid criteria were established early. These included manner of presentation, size of illustrations, as well as style of headings, subheadings and legends. The authors were encouraged to keep the text at a minimum, freeing as much space as possible for large illustrations and meaningful legends. The "Atlas" is to be just that, an "atlas," not a series of "texts." The authors were urged, also, to keep the bibliography brief.

The selection of suitable authors for the "Atlas" was extremely difficult, and to a degree invidious. For the final choice, the Editor-in-Chief accepts full responsibility. It is but fair to record, however, that his Editorial Advisory Committee accepted his recommendations. The format of the "Atlas," too, was the choice of the Editor-in-Chief, again with the concurrence of his advisory group. Should the "Atlas of Tumor Radiology" fall short of its goals, the fault will lie with the Editor-in-Chief alone; his Editorial Advisory Committee was selfless in its dedication to the purposes of the "Atlas," rendering invaluable advice and guidance whenever asked to do so.

As medical knowledge expands, medical concepts change. In medicine, the written word considered true today may not be so tomorrow. The text of the "Atlas," considered true today, therefore may not be true tomorrow. What may not change, what may ever remain true, may be the illustrations of the "Atlas of Tumor Radiology." Their legends may change as our conceptual levels advance. But the validity of the roentgen findings there recorded should endure. Thus, if the fidelity with which the roentgenograms have been reproduced is of superior order, the illustrations in the "Atlas" should long serve as sources for reference no matter what revisions of the text become necessary with advancing medical knowledge.

ACKNOWLEDGMENTS

The American College of Radiology, its Commission on Education, the Editorial Advisory Committee, the authors and the Editor-in-Chief wish to acknowledge their grateful appreciation:

1. For the grants-in-aid so willingly and repeatedly provided by The American Cancer Society, The American Roentgen Ray Society, The Cancer Control Program, National Center for Chronic Disease Control (USPHS Grant No. 59481), The James Picker Foundation, and the Radiological Society of North America.

2. For the superb glossy print reproductions provided by the Radiography Markets Division, Eastman Kodak Company. Special mention must be made of the sustained interest of Mr. John Fink, its Assistant Vice-President and General Manager. We applaud particularly Mr. William S. Cornwell, Technical Associate and Editor Emeritus of Kodak's *Medical Radiography and Photography,* as well as his associates, Mr. Charles C. Heckman and Mr. David Edwards and others in the Photo Service Division whose expertise provided the "Atlas" with its incomparable photographic reproductions.

3. To Year Book Medical Publishers, for their personal involvement with and judicious guidance in the many problems of publication. There

were occasions when the publisher questioned the quality of certain illustrations. Almost always the judgment of the authors and the Editor-in-Chief prevailed because of the importance of the original roentgenograms and the singular fidelity of their reproduction.

4. To the Associate Editors, particularly Mrs. Anabel I. Janssen, whose talents lightened the burden of the Editor-in-Chief and helped establish the style of presentation of the material.

5. To the Staff of the American College of Radiology, especially Messrs. William C. Stronach, Otha Linton, Keith Gundlach and William Melton, for continued conceptual and administrative efforts of unusual competence.

This volume could never have been completed had our authors been less than totally dedicated. One could justly say of both men "no-wher so bisy a man as he ther was" (The Canterbury Tales).

In a time of national economic crisis our senior author shouldered much of his institution's administrative responsibilities in addition to his clinical load. Coincidentally, as President of both the American College of Radiology and the Radiological Society of North America he carried commitments enough to more than tax the ordinary. It was not any easier for our junior author. He, too, was strained to surmount heavy responsibilities imposed by a major shift in his academic scene. With equal selflessness he met the challenge.

When this work was first planned barium was the contrast agent universally used to identify abdominal masses. The delicate vascular nuances now recognized as hallmarks of abdominal disease were still to be fully appreciated. By the time this volume was completed, however, abdominal angiography had become commonplace. We predict that angiography, too, will be supplanted, in part at least, by noninvasive techniques now coming into focus. Advances in ultrasound and the EMI-scanner lend substance to this prediction.

PHILIP J. HODES
Editor-in-Chief

Emeritus Professor of Radiology,
Thomas Jefferson University, Philadelphia
Professor of Radiology, University of Miami School
of Medicine, Miami, Florida

Editorial Advisory Committee

HARRY L. BERMAN   VINCENT P. COLLINS   E. RICHARD KING
LEO G. RIGLER   PHILIP RUBIN

Author's Preface

THE DIAGNOSIS of tumors of the accessory digestive organs encompasses virtually all branches of the discipline of radiology. This has resulted in a high level of accuracy in the diagnosis of lesions of these organs but at once has compounded the problem of assembling material for a tumor atlas to illustrate the vast number of lesions and techniques attending the diagnosis of these lesions. In addition, several of the modalities of diagnosis were in a rapidly developing state when this task was first begun. This, of course, resulted in a need to update material constantly. The tried and true conventional techniques, such as the upper gastrointestinal examination, were not a problem, since our own files contained abundant illustrations. However, in the case of angiography it was necessary to draw upon the resources of others in order to depict the broad spectrum of diagnosis more completely. We naturally turned to one of the pioneers in angiography, Dr. Stanley Baum, who fortunately, by this time, was on the Boston scene at the Massachusetts General Hospital. The generosity of Dr. Baum, as the reader will note, has resulted in many superb angiographic illustrations of tumors. In a similar fashion we were fortunate in being able to draw upon the resources of a pioneer in nuclear scanning of the pancreas, Dr. Antonio Rodriguez-Antunez of the Cleveland Clinic Foundation. His cooperation has also resulted in a significant contribution to this volume. Sialography, a highly specialized diagnostic procedure in which few have become accomplished, presented a particular problem. Since our own experience in this area was severely limited, we were fortunate in having Dr. Alexander S. Macmillan, Jr., of the Massachusetts Eye and Ear Infirmary as our neighbor and colleague. Doctor Macmillan freely made available his collection of sialograms. In addition, he contributed the basis of Part 5 on sialography.

As with every other form of human endeavor, progress leads to abandonment or, at the least, decreased use of established procedures. Thus it has been with the diagnosis of diseases of the liver and pancreas. While the standard upper gastrointestinal examination with barium remains our mainstay of diagnosis, especially with respect to the pancreas, angiography has led to a higher degree of specificity. Without its use many lesions would remain undiagnosed until surgical exploration and frequently after as well. The development of the flexible fiberoptic endoscope that permits cannulation of the pancreatic ducts and injection of contrast media has led to an even higher level of diagnostic accuracy. We also find ourselves less dependent upon intravenous and percutaneous cholangiography in the diagnosis of diseases of the bile ducts and the liver now that the common bile duct can be injected by way of the endoscope.

While we may regret the passing of some time-honored techniques, we must be mindful that all of the previous knowledge gained through the use of these techniques is immediately applicable to the radiographs that are obtained through the use of endoscopic cannulation techniques.

I have been fortunate in being associated with an unusually large collection of talented and helpful associates, without whom this volume would never have been completed. I refer specifically to my colleagues in the Department of Diagnostic Radiology of the Lahey Clinic Foundation whose forbearance, help and support are most appreciated. My secretary, Miss Carolyn Betts, was most helpful in the selection and preparation of the illustrations and text. Miss Celia Medeiros, Managing Editor of the *Lahey Clinic Foundation Bulletin,* assumed the responsibility for the organization and editorial aspects of the project. Her skill and cooperation were the catalysts that ultimately brought this work to fruition.

Having accepted the responsibility for the production of this volume and other obligations simultaneously, I eventually came to the realization that without a coauthor the book could not be completed in the allotted period of time. At this time we were most fortunate in having with us in the Department of Diagnostic Radiology a remarkably energetic and capable resident, Dr. Austin P. O'Keeffe. Although cast in the role of junior author, his contribution has been a major one, and I cannot be too complimentary concerning his skill as a writer and organizer. I am deeply appreciative of his work, for without it this volume would have been long delayed and its quality would have been diminished.

Mr. William S. Cornwell of the Eastman Kodak Company merits special recognition. We were fortunate in being able to benefit from his mastery of photographic skills. Through his unrelenting efforts along with the staff of the Eastman Kodak Company, they have produced unsurpassable illustrations of the highest quality throughout the book.

The efforts of Dr. Philip J. Hodes cannot go unrecognized. The role of Editor-in-Chief is definitely more difficult than the reader, who may never have been associated with a task of this type, can ever appreciate. Doctor Hodes has found it necessary to work closely with the authors on their illustrations and text. These, however, have been among the least of his problems. It has been necessary for him, in at least this instance, to serve as "whip," to cajole, bribe, heckle and, yea, at times even to threaten in order to achieve production. To Dr. Hodes I give my heartfelt thanks for his patience, perseverance and helpfulness. We are grateful to you, Dr. Hodes.

ROBERT E. WISE

PART 1

The Pancreas, 1

PART 2

The Liver, 119

PART 3

The Extrahepatic Bile Ducts, 227

PART 4

The Gallbladder, 275

PART 5

The Salivary Glands, 303

Index, 329

PART 1

The Pancreas

Pancreatic Tumor Characteristics

ADENOCARCINOMA

Carcinoma of the pancreas comprises 1.75–2% of all malignant tumors. Review of mortality trends indicates that the mortality rate from pancreatic carcinoma has distinctly increased in the past 40 years and the relative frequency continues to increase. The disease affects two to three times as many men as women, and the average age at the time of diagnosis is 57 years.

Two diseases of the pancreas most frequently found in association with carcinoma are chronic pancreatitis and diabetes mellitus. Although pancreatic carcinomas are known to occur in patients with chronic pancreatitis, this occurs only in the exceptional case and may well be coincidental. Carbohydrate metabolism is disturbed in many patients with carcinoma of the pancreas. However, impaired glucose tolerance may be found with any carcinoma or debilitating disease. It is not proved that true genetic diabetes mellitus predisposes to carcinoma of the pancreas.

PATHOLOGY.—Two out of three carcinomas of the pancreas occupy mainly the head of the gland; the rest are divided equally between the body and the tail of the pancreas. A diffuse nodular type occurs, and this is difficult to distinguish from normal parenchyma by macroscopic examination. Most pancreatic adenocarcinomas are large, approximately 6 cm in diameter, when found at operation or at autopsy. The carcinoma may be sharply demarcated or permeative. Many extend to contiguous areas of the pancreas, and it is frequently difficult to determine the site of origin.

Pancreatic carcinomas are solid, scirrhous tumors. Approximately 80% are of ductal cell origin, and the remainder arise from the acinar cell. They contain varying degrees of desmoplastic tissue, epithelial cells, acinar structures and mucin. Anaplasia is commonly found.

Pancreatic adenocarcinoma exhibits rapid growth and spread by invasion and direct extension. The tumor also spreads by way of the lymphatic and the blood vessels and by invasion of perineural tissue. This propensity for involvement of nerves may account for the severe pain associated with pancreatic adenocarcinoma. Metastases are commonly found in the regional lymph nodes and the liver and may also occur in the peritoneum and the lung. The principal mode of spread, direct extension, frequently involves the tissues around the major mesenteric vessels, making the tumor unresectable.

CLINICAL FEATURES.—Upper abdominal pain is the first symptom in three out of four patients. The pain is severe and unremitting. It may be felt mainly in the epigastrium but is frequently transmitted to the back and may be more severe in the supine position. Jaundice is present in two-thirds of patients when initially examined. Weight loss, anorexia, nausea and vomiting are other frequent presenting symptoms. Abnormal bowel function is also common. Thromboembolic phenomena and acute gastrointestinal hemorrhage occasionally complicate carcinoma of the pancreas.

RADIOGRAPHIC FINDINGS.—The routine radiograph of the abdomen is usually not helpful in making the diagnosis of pancreatic carcinoma. Displacement of pancreatic calcification, whose position has changed from previous roentgen observations, is sometimes noted. A soft tissue mass in the region of the pancreas is infrequently seen. Displacement of air-filled viscera may be evident. Tumor impression on viscera is an important factor in evaluating the presence of a retroperitoneal mass. Barium gastrointestinal studies give indirect clues to pancreatic neoplasm when there is evidence of pressure, displacement, deformity, rigidity or ulceration involving the stomach, duodenum or transverse colon in areas contiguous to the pancreas. Invasion of these viscera may simulate intraluminal tumors. The region of the gastrointestinal tract that is involved depends on the site of the tumor within the pancreas. Changes occurring in the stomach include anterior or lateral displacement, pressure defects on the posterior wall and elevation of the antrum. In the duodenum the loop may be enlarged; there may be depression of the ligament of Treitz, and anterior displacement of the proximal duodenum is frequently noted.

A double contour on the medial aspect of the duodenal loop is sometimes seen. Mucosal folds may be flattened, nodular, rigid, thick or divergent. The reverse-figure-3 sign is occasionally evident on the medial aspect of the second portion of the duodenum. Hypotonic duodenography is useful in the diagnosis of pancreatic carcinomas that affect the duodenal loop, as the previously mentioned signs of involvement of the duodenum will be more in evidence on this examination. Diverticula occurring within the duodenal loop may be distorted by pancreatic adenocarcinoma.

The gallbladder is frequently enlarged in association with adenocarcinoma of the pancreas as a result of partial or complete obstruction of the common bile duct when the tumor occurs in the head of the pancreas or extends to the head of the pancreas. This obstruction and resultant increased serum bilirubin levels prevent opacification of the bile ducts and gallbladder by oral or intravenous contrast agents. The *percutaneous transhepatic chol-*

angiogram is then useful in demonstrating the dilated biliary tract and the outline of the common bile duct at the site of obstruction.

Pancreatic *radionuclide studies* using selenium-75 (selenomethionine) have some application in screening patients for pancreatic masses. A normal scan has a high degree of reliability; on an abnormal scan it is usually not possible to distinguish pathologic entities. A false-negative rate of 12% is to be anticipated. Sequential imaging with the gamma camera is used principally to establish normality or abnormality of the gland, without attempting an etiologic diagnosis. Because of the uptake of selenomethionine by the liver also, there is occasionally some difficulty in differentiating between the pancreas and the liver. If the liver is first scanned using technetium sulfur colloid and this image is later subtracted either photographically or by computer from the selenomethionine scan, a less confusing and diagnostically more accurate study is achieved.

Angiographic studies of the pancreas are reliable and accurate in the detection of pancreatic carcinoma. A diagnostic accuracy rate of 86% is reported for adenocarcinoma of the pancreas in one series. In particular, angiography has been found to be reliable in differentiating adenocarcinoma from pancreatitis. The procedure is useful in evaluating operability as well. The principal diagnostic features are:

1. Arterial stenosis or encasement. This usually involves a short segment of artery and is the result of tumor surrounding and stenosing the arterial wall. The vessel border may be smooth, serrated or serpiginous.

2. Tumor vessels and neovascularity. Adenocarcinomas are usually avascular on angiographic study; that is, although the tumor must have vascularization, the vessel size is below the resolution possible on angiographic study. Tumor vessels (vessels that exhibit changes, such as an abnormal course and branching pattern, abnormal dilatations and narrowings, and haphazard arrangement) are frequently demonstrated. Displacement of major vessels, such as the splenic artery, the proper hepatic artery or the gastroduodenal artery, is sometimes seen. Because of the poor vascularity of the tumor bed, a tumor stain is rarely seen.

3. Vascular obstruction. Major branch arteries supplying the pancreas may be obstructed. The gastroduodenal artery or the dorsal pancreatic artery may have an abrupt termination. Venous obstruction, particularly obstruction of the splenic vein, is a common angiographic finding in pancreatic carcinoma. Involvement of veins may be exhibited by displacement, narrowing, irregularity of the vessel wall, opacification of collateral veins and complete obstruction. The superior mesenteric and the portal veins are also commonly involved by these processes. In one study, venous involve-

ment was the most common angiographic finding in pancreatic adenocarcinoma. Concurrent with the diagnosis of pancreatic adenocarcinoma, evidence of hepatic metastases or enlarged gallbladder may be found on celiac angiography. Although vessel narrowing is found in pancreatitis also, a beaded appearance is frequently present, and the lesion may be hypervascular. Small artery occlusions also occur in pancreatitis. The narrowing may be caused by atherosclerosis; involvement of a major artery with normal veins is frequently found in this entity, as a point of differentiation.

Cystadenoma and Cystadenocarcinoma

Cystadenoma and cystadenocarcinoma are extremely rare, but it is important to distinguish them from adenocarcinoma and pseudocyst of the pancreas. The benign cystadenoma is more common than cystadenocarcinoma. Both occur much more frequently in women than in men in a ratio of 8:1, and their greatest frequency is in the fifth, sixth and seventh decades. They are more frequently located in the body and tail of the pancreas than in the head. In a small percentage of cases multiple areas of involvement are found.

These tumors arise de novo from aberrant proliferation of pancreatic duct epithelium with numerous small cystic spaces that do not communicate with the normal ductal system and are situated in abundant myxomatous stroma. Papillary epithelial projections are also found within these cystic spaces. The stroma and the papillary projections are vascular.

Slow progress of nonspecific symptoms of an abdominal retroperitoneal mass is the usual first manifestation. Epigastric pain, back pain, disturbance of bowel function, nausea, vomiting and jaundice are the most common presenting symptoms. An upper abdominal lesion is rarely palpable. There is a 10–20% coincidence of diabetes mellitus.

RADIOGRAPHIC FINDINGS.—Calcification with a distinctive radiating pattern is present in 10% of cystadenomas of the pancreas. Otherwise, the results of routine studies, biliary tract radiographs and gastrointestinal tract studies do not differ significantly from those of adenocarcinoma of the pancreas.

On angiographic studies of the pancreas, the cystadenomas and cystadenocarcinomas are extremely vascular because of the richly vascular stroma and the papillary projections. Numerous tumor vessels and tumor neovascularity are seen, and a dense tumor stain occurs during a late phase of the injection. Avascular areas corresponding to these cysts are scattered throughout the tumor opacification. The feeding arteries are often enlarged as a result of the increased demands of the tumor. Displacement of neighboring large vessels occurs because of the tumor mass, but tumor encasement and obstruction of vessels is not as frequent as in adenocarcinoma.

ISLET CELL TUMORS

The endocrine secretory function of the pancreas is accomplished by the islets of Langerhans. These are clusters of small round cells containing acidophilic or basophilic granules that are diffusely scattered within the pancreas, each islet being about four times the size of an exocrine acinus. Depending on the staining properties of their granules, the cells are named alpha or beta cells.

Approximately 20% of islet cell tumors are hormonally active. The functional tumors of the beta cell type are called *insulinomas*. These are responsible for excessive secretion of insulin resulting in hypoglycemic attacks. The symptom complex associated with this disease entity is termed Whipple's triad and consists of (1) attacks of spontaneous hypoglycemia in association with both central and sympathetic nervous system symptoms including sweating, tachycardia, stupor and unconsciousness; (2) a blood sugar level repeatedly less than 50 mg/100 ml, and (3) an instant reversal of symptoms with the intravenous administration of glucose. Twelve per cent of beta cell tumors are carcinomas, and approximately two-thirds of these tumors occur in the body and tail of the pancreas. In 10–15% of cases multiple tumors are found. Diffuse hyperplasia of all the islets may occur. Islet cell tumors may be found in ectopic foci of pancreatic tissue.

The islet cell *adenoma* is a small tumor, usually less than 2 cm in diameter, and is well encapsulated. It is most commonly found in the 30- to 60-year age group with its incidence being the same for both sexes.

The *Zollinger-Ellison syndrome* is ascribed to functional tumors of the alpha cells of the islet of Langerhans. These cells have some similarity to the argentaffin cells found in carcinoid tumors. The islet cell tumors responsible for the Zollinger-Ellison syndrome are believed to secrete gastrin. The syndrome is characterized by gastric hypersecretion, recurrent peptic ulceration and diarrhea. Response to the 12-hour overnight hydrochloric acid secretion test is greater than 100 mEq of free acid. The ulceration is refractory to treatment, may be multiple and may involve an unusual site, such as the jejunum. Diarrhea may be severe and debilitating. The tumors are frequently multiple and may be accompanied by adenomas in other endocrine organs (Wermer's syndrome). Malignant tumors responsible for the Zollinger-Ellison syndrome are found in 61% of patients. These alpha cell tumors occur more frequently in the head and body of the pancreas, grow to a larger size, are more frequently malignant, and 44% have metastasized when diagnosed. They metastasize mainly to the pancreaticoduodenal lymph nodes and the liver; progression of the metastasis is slow.

RADIOGRAPHIC FINDINGS.—The small size of islet cell tumors plus their frequency of occurrence distally in the pancreas make them relatively inaccessible radiographically. In the Zollinger-Ellison syndrome, the upper gastrointestinal examination may show evidence of ulceration, hypermotility and thickened edematous mucosa in the stomach and small bowel. Atony and dilatation of the stomach and duodenum may be observed.

Angiographic evaluation of the pancreas by selective arteriography of the celiac and superior mesenteric arteries successfully demonstrates an islet cell tumor in the majority of instances. Numerous small vessels are demonstrated in the arterial phase, and a dense, well-circumscribed blush is noted in the capillary phase. The vessels feeding the tumor are usually increased in size. Alpha cell tumors are generally larger than insulinomas. Tumors less than 1 cm are extremely difficult to demonstrate angiographically. The benign or malignant state of the tumor usually cannot be determined from the angiographic appearance itself. However, a large, irregular-appearing tumor is suggestive of malignancy, and observation of vascular metastases within the liver or elsewhere in the abdomen confirms the diagnosis.

CYSTS

This category of pancreatic tumors includes pseudocysts, which are numerically and clinically the most important cystic lesions of the pancreas. Cystic neoplasms have been described previously and will not be discussed here. Other pancreatic cysts are retention cysts, congenital cysts and parasitic cysts. Pancreatitis is by far the most frequent cause of pancreatic pseudocysts.

Pseudocysts are generally large tumors containing more than 100 ml of fluid which may be clear or turbid but at times may be mucoid. Occasionally, necrotic hemorrhagic tissue is contained within the cystic cavity. The cyst does not have an epithelial lining and is limited by a fibrous wall. Pseudocysts may be intrapancreatic but generally are found adjacent to, but outside, the pancreas. The most common site is within the lesser sac posterior to the midbody of the stomach. Extension into the mediastinum or inferiorly into the pelvis is known to occur.

Pancreatic pseudocysts are more frequent in men than in women in a ratio of 2:1, and this reflects the increased incidence of pancreatitis in men. A history of pancreatitis or abdominal trauma is available in most instances, although occasionally no attributable cause is found. At least 75% of patients have a palpable upper abdominal mass. Associated symptoms are pain, nausea and vomiting, fever, diarrhea, abdominal tenderness and weight loss. Jaundice is evident in 10% of patients; the serum amylase value is increased in 50% of patients.

RADIOGRAPHIC FINDINGS.—The most frequent findings on routine radiographic examination are soft tissue mass, pancreatic calcification and bowel gas displacement. Curvilinear cyst wall calcification is rarely demonstrated. The most common radiographic finding with pancreatic pseudocyst is anterior displacement of the stomach, and for this reason a lateral view of the stomach should be obtained when pancreatic pseudocyst is suspected. Displacement and evidence of extrinsic pressure may involve the duodenum, colon, kidney or common bile duct, depending on the position of the pseudocyst. On arteriographic examination by selective celiac and superior mesenteric injections, the most frequent and characteristic finding is smooth vessel displacement by a large avascular mass in the pancreatic region.

Retention cysts of the pancreas result from chronic pancreatitis or trauma due to obstruction of a major pancreatic duct. The cysts are intrapancreatic, are usually multiple and are generally much smaller than pseudocysts. Congenital cysts are caused by cystic dilatation of aberrant components of pancreatic ducts that do not communicate with the normal ductal system. These cysts are generally multiple and of small diameter. They may be associated with mucoviscoidosis and congenital cysts of the kidneys and liver.

Hydatid cysts are known to occur within the pancreas but are extremely rare.

SUGGESTED READING

Abrams, R. M., et al.: Angiographic studies of benign and malignant cystadenomas of the pancreas, Radiology 89:1028–1032, December, 1967.

Baum, S., et al.: Clinical application of selective celiac and superior mesenteric arteriography, Radiology 84:279–295, February, 1965.

Beranbaum, S. L.: Carcinoma of the pancreas: A bi-directional roentgen approach, Am. J. Roentgenol. 96:447–467, February, 1966.

Bieber, W. P., and Albo, R. J.: Cystadenoma of the pancreas: Its arteriographic diagnosis, Radiology 80:776–778, May, 1963.

Bookstein, J. J., Reuter, S. R., and Martel, W.: Angiographic evaluation of pancreatic carcinoma, Radiology 93:757–764, October, 1969.

Buranasiri, S., and Baum, S.: The significance of the venous phase of celiac and superior mesenteric arteriography in evaluating pancreatic carcinoma, Radiology 102:11–20, January, 1972.

Gray, R. K., Rösch, J., and Grollman, J. H., Jr.: Arteriography in the diagnosis of islet-cell tumors, Radiology 97:39–44, October, 1970.

Korobkin, M. T., Palubinskas, A. J., and Glickman, M. G.: Pitfalls in arteriography of islet-cell tumors of the pancreas, Radiology 100:319–328, August, 1971.

Landman, S., Polcyn, R. E., and Gottschalk, A.: Pancreas imaging—is it worth it? Radiology 100:631–636, September, 1971.

Mani, J. R., Zboralske, F. F., and Margulis, A. R.: Carcinoma of the body and tail of the pancreas, Am. J. Roentgenol. 96:429–466, February, 1966.

Nebesar, R. A., and Pollard, J. J.: A critical evaluation of selective celiac and superior mesenteric angiography in the diagnosis of pancreatic diseases, particularly malignant tumor: Facts and "artefacts," Radiology 89:1017–1027, December, 1967.

Reuter, S. R.: Superselective pancreatic angiography, Radiology 92:74–85, January, 1969.

Robins, J. M., *et al.*: Selective angiography in localizing islet-cell tumors of the pancreas: A further appraisal, Radiology 106:525–528, March, 1973.

Shockman, A. T., and Marasco, J. A.: Pseudocysts of the pancreas, Am. J. Roentgenol. 101:628–638, November, 1967.

Zboralske, F. F., and Amberg, J. R.: Detection of the Zollinger-Ellison syndrome: The radiologist's responsibility, Am. J. Roentgenol. 104:529–543, November, 1968.

Figure 1.—Adenocarcinoma of the pancreas.

A 65-year-old man had persistent upper abdominal pain that was initially diagnosed as a gastric ulcer that did not respond to treatment. At operation, a pancreatic adenocarcinoma, measuring 5 cm in diameter, was found involving the midbody of the pancreas.

A, upper gastrointestinal radiograph, lateral projection: Evidence of extrinsic pressure is seen on the posterior aspect of the distal body and antrum of the stomach (**arrowheads**). The stomach wall is unevenly indented, but there is no evidence of invasion of the wall.

(*Continued.*)

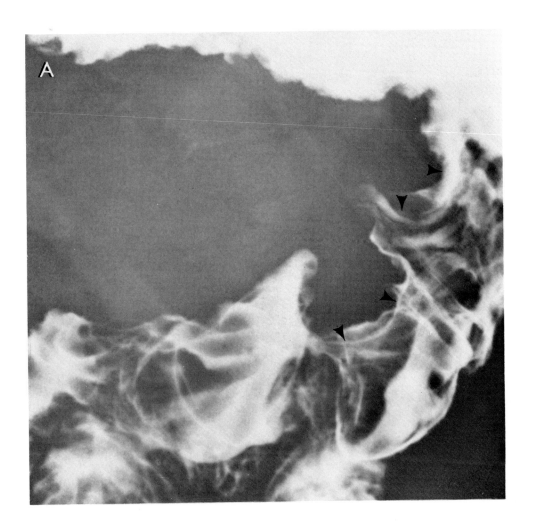

Figure 1 · Adenocarcinoma

Figure 1 (cont.).—Adenocarcinoma of the pancreas.

B, selective celiac angiogram, early arterial phase: Note irregular narrowing of the proximal splenic artery (**arrowhead**). The gastroduodenal and dorsal pancreatic arteries are not visible because of complete obstruction at their origins. The branches of the cystic artery are separated, indicating enlargement of the gallbladder.

C, selective celiac angiogram, venous phase: The splenic vein is completely obstructed near its origin. A markedly tortuous collateral vein (**arrows**) is well opacified; the portal vein is not yet opacified. The gallbladder is greatly enlarged (**arrowheads**).

Comment: This case exhibits classic radiographic findings on upper gastrointestinal and angiographic examinations. A sizable pancreatic tumor arising in the midportion of the pancreas frequently indents the posterior wall of the stomach and may invade it. Arterial encasement and obstruction, venous obstruction with collateral flow, distended gallbladder and a mottled hepatogram indicating hepatic metastasis are angiographic findings commonly found with adenocarcinoma of the pancreas.

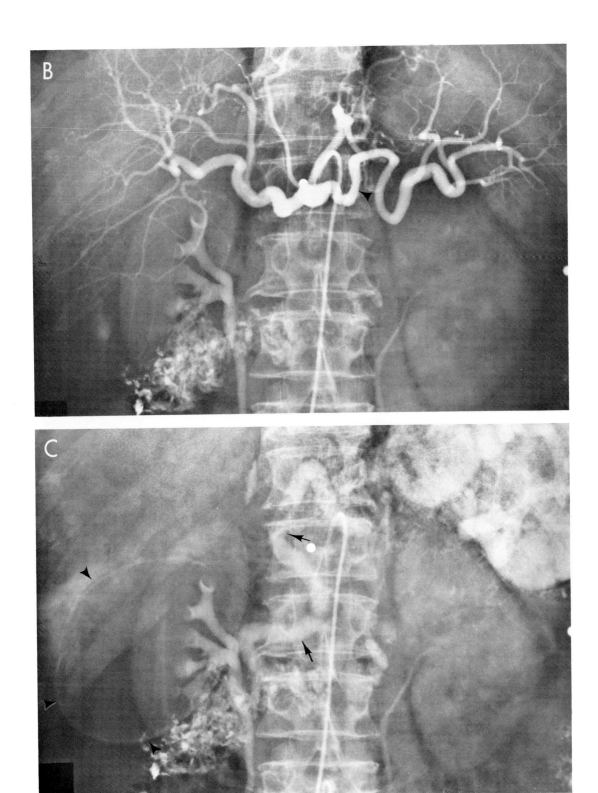

Figure 1 · Adenocarcinoma

Figure 2.—Adenocarcinoma of the pancreas.

A 47-year-old woman had a history of intermittent abdominal distress. Nausea, vomiting and marked weight loss were the major symptoms before hospitalization. The liver was markedly enlarged, nodular and tender.

A, intravenous cholangiogram, right posterior oblique projection: The common bile duct is delineated, and a long, tapered narrowing of its distal portion within the head of the pancreas is seen (**arrowheads**). Above this segment the common hepatic duct is dilated.

B, selective celiac angiogram, arterial phase: Note short segment of marked narrowing of the proximal common hepatic artery typical of tumor encasement (**arrowheads**). The intrahepatic arteries are stretched due to liver enlargement. The gastroduodenal artery is not visible during this injection because it is obstructed.

(*Continued.*)

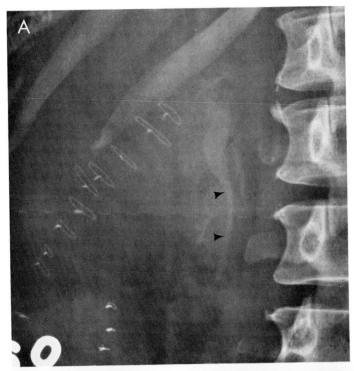
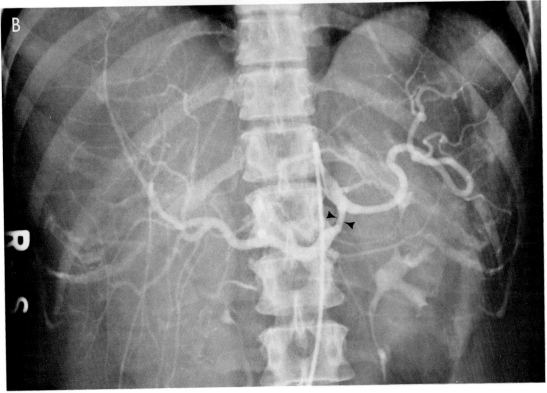

Figure 2 · Adenocarcinoma / 17

Figure 2 (cont.).—Adenocarcinoma of the pancreas.

C, selective superior mesenteric angiogram, arterial phase: There is retrograde filling of the gastroduodenal artery, the proper hepatic artery and the hepatic vessels by way of the pancreatic arcade. This is indicative of decreased flow in the stenosed hepatic artery.

D, selective celiac angiogram, venous phase: The splenic vein is completely obstructed near the midline (**arrowheads**). The liver is greatly enlarged, and numerous lucencies within the hepatogram indicate metastasis (**arrows**).

Comment: Adenocarcinoma of the pancreas arising in or extending to the head of the pancreas may cause either partial or complete obstruction of the intrapancreatic portion of the common bile duct.

In this instance, the long, tapered narrowing of the bile duct was caused by extrinsic pressure from a carcinoma of the head of the pancreas. The obstruction was incomplete, and jaundice had not yet developed.

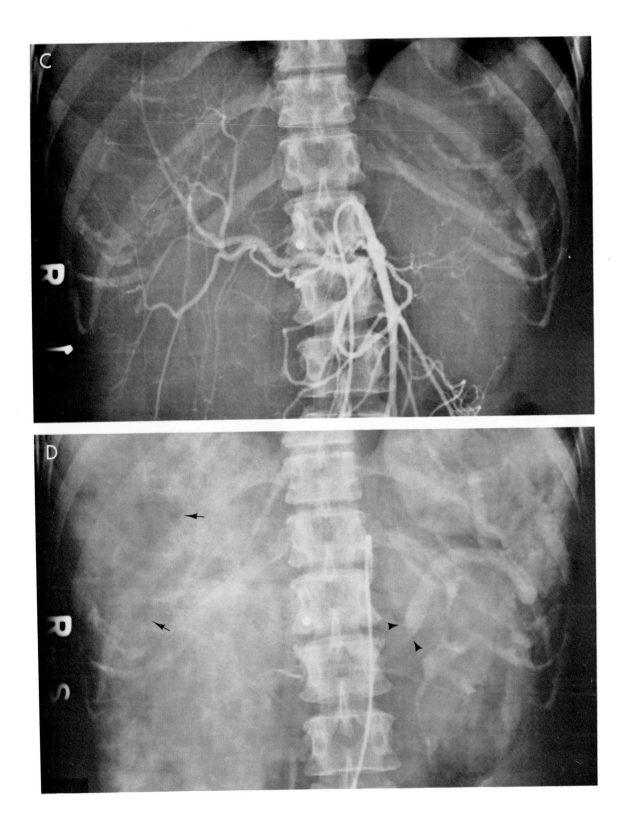

Figure 2 · Adenocarcinoma / 19

Figure 3.—Adenocarcinoma of the pancreas.

A, selective superior mesenteric angiogram, arterial phase, right posterior oblique projection: Note minimal irregularity and narrowing of the proximal right hepatic artery approximately 2 cm from its origin from the superior mesenteric artery (**arrowhead**).

B, selective superior mesenteric angiogram, venous phase, right posterior oblique projection: The superior mesenteric vein is completely obstructed at the level of the lower border of the second lumbar vertebra. Blood flow is borne by the large collateral vein (**arrowheads**) which communicates with the portal vein. There is good opacification of the intrahepatic portal venous system.

(*Continued.*)

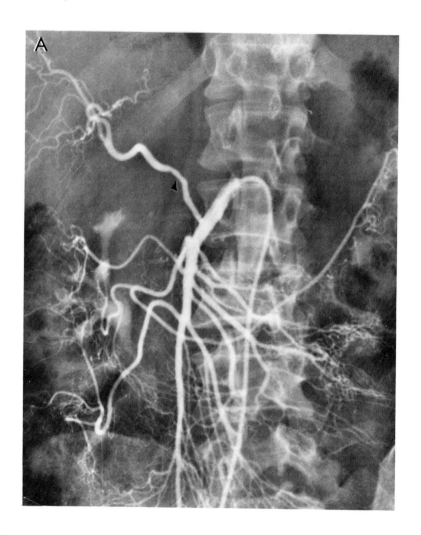

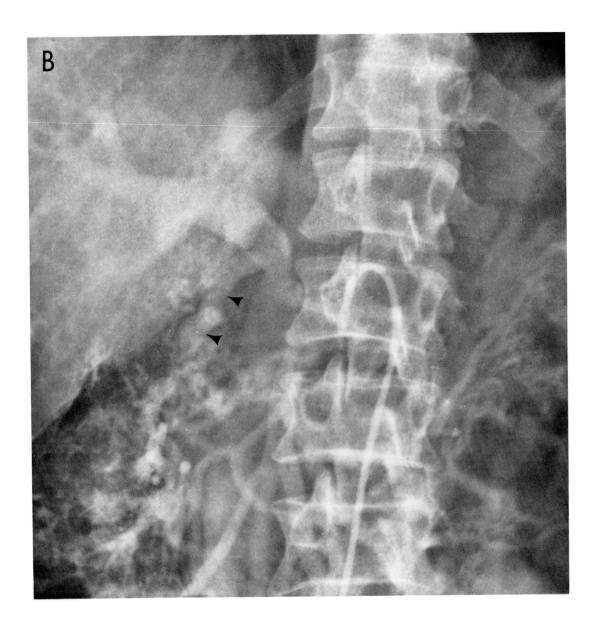

Figure 3 · Adenocarcinoma / 21

Figure 3 (cont.).—Adenocarcinoma of the pancreas.

C, selective superior mesenteric angiogram, venous phase, subtraction radiograph in right posterior oblique projection: By use of the subtraction technique the area of complete obstruction of the superior mesenteric vein is better defined (**arrows**), as are the numerous large collateral veins.

Comment: Dr. Buranasiri and Dr. Baum reported venous involvement in 46 of 47 cases of carcinoma of the pancreas and consider this to be the most constant angiographic finding in this condition.

Technical note: When conventional methods fail to demonstrate adequate venous filling, tolazoline hydrochloride (Priscoline), 25–50 mg, may be injected directly into the superior mesenteric artery immediately before the injection of contrast material. This frequently results in a greatly improved venous phase of the arteriogram.

Figure 3, courtesy of Dr. S. Baum, Massachusetts General Hospital, Boston; *A* and *B* reprinted from Buranasiri, S., and Baum, S.: Radiology 102:12, January, 1972.

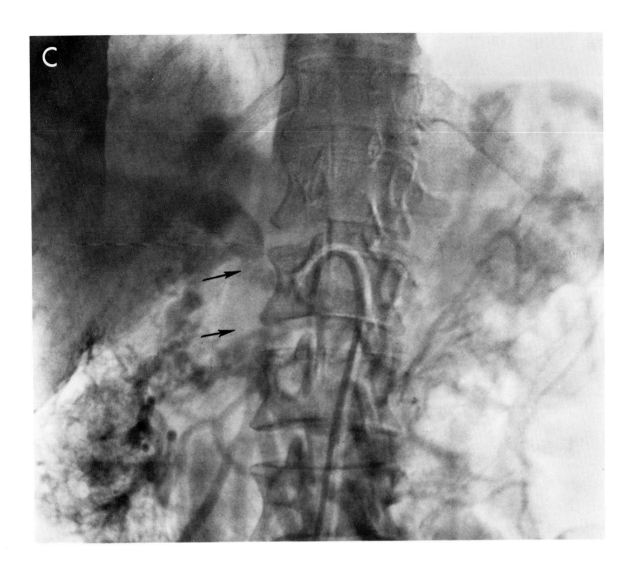

Figure 4.—Adenocarcinoma of the pancreas.

A 62-year-old woman complained of epigastric pain and a 14-lb weight loss. The upper abdomen was tender to palpation.

A, selective celiac angiogram, arterial phase: The gastroduodenal artery is completely obstructed approximately 2 cm distal to its origin from the common hepatic artery (**arrowhead**).

B, selective superior mesenteric angiogram, arterial phase: The inferior pancreaticoduodenal artery is narrowed near its origin from the superior mesenteric artery (**arrow**). Numerous small tumor vessels and a minimal tumor stain are present in the region of the head of the pancreas (**arrowheads**).

Comment: Rich tumor vascularity and stain are not often demonstrated in pancreatic adenocarcinoma, for pancreatic adenocarcinoma is not a richly vascular tumor. The tumor neovasculature can frequently be resolved by using superselective and serial magnification techniques.

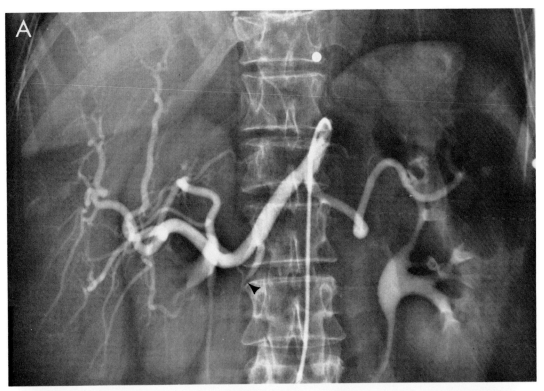
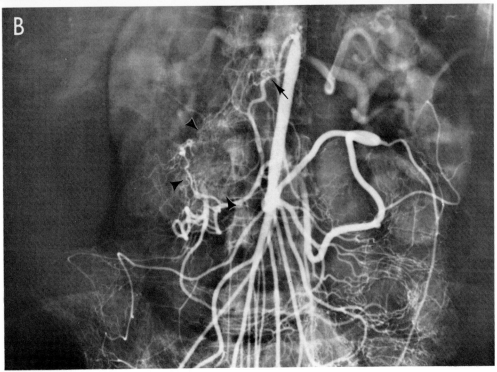

Figure 4 · Adenocarcinoma / 25

Figure 5.—Adenocarcinoma of the pancreas.

A 60-year-old man had a one-month history of epigastric pain radiating to the back, 30-lb weight loss over a longer period and jaundice for two weeks. The liver was enlarged 4 cm below the right costal margin. At operation, an 8-cm tumor was found involving the head and proximal body of the pancreas.

A, upper gastrointestinal radiograph, right anterior oblique spot film: Note distortion of the duodenal bulb with evidence of extrinsic pressure on the bulb (**arrowheads**) but no definite evidence of invasion of the duodenal wall. The extrinsic mass appears to lie posteriorly.

B, superselective gastroduodenal arteriogram, arterial phase: There is irregular narrowing of the proximal gastroduodenal artery, and the artery is displaced laterally (**arrowheads**). The anterior pancreatic duodenal arcade appears to be widened and rigid.

C, selective celiac angiogram, venous phase: The splenic vein is completely obstructed. A large collateral vein traverses the abdomen.

Comment: Direct injection into an artery supplying the pancreas enhances the delineation of the intrapancreatic vasculature. However, injection must be made into at least two arteries supplying the pancreas. Intrapancreatic arterial communications do exist but are inadequate for total pancreatic opacification with one superselective injection.

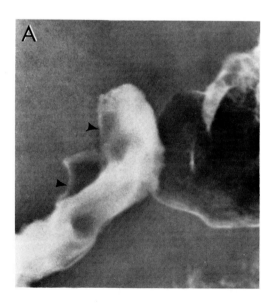

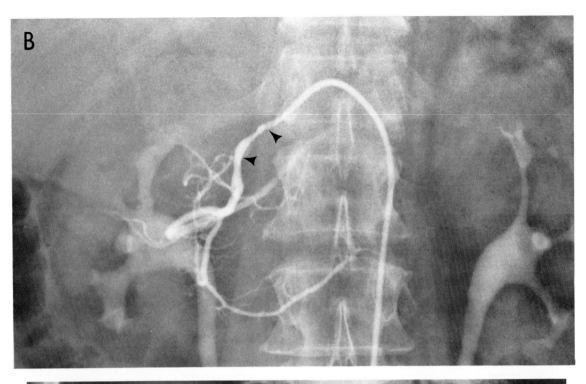

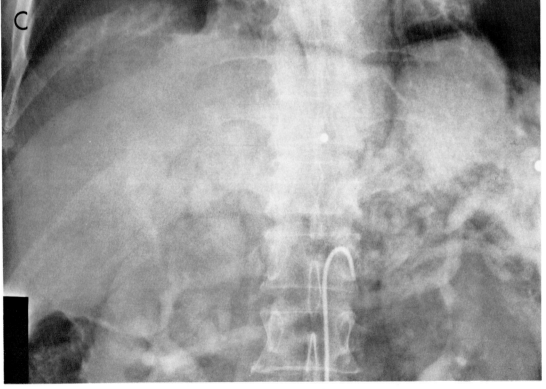

Figure 5 · Adenocarcinoma / 27

Figure 6.—Adenocarcinoma of the pancreas.

This patient had had weight loss and weakness for four months and complained of epigastric pain radiating to the back. There was recent onset of jaundice and diarrhea.

A, right anterior oblique spot film of the duodenal loop: The duodenum is markedly narrowed at the junction of the second and third portions (**arrows**). The mucosa in this area appears to be invaded, stretched and ulcerated.

(*Continued.*)

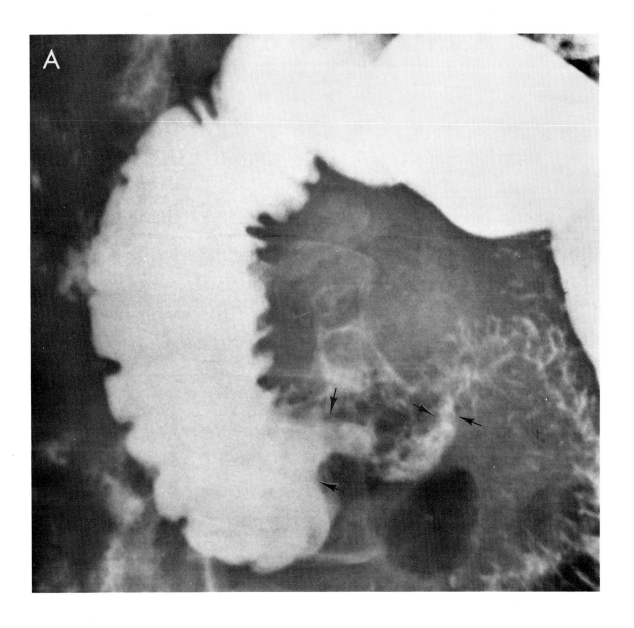

Figure 6 · Adenocarcinoma / 29

Figure 6 (cont.).—Adenocarcinoma of the pancreas.

B, selective celiac angiogram, arterial phase: Note tortuosity, narrowing and dilatation of the small pancreatic vessels arising from the anterior pancreaticoduodenal branch of the gastroduodenal artery (**arrowheads**) and irregularity of the wall of the common hepatic artery near the origin of the gastroduodenal artery. The gastroduodenal artery itself is minimally narrowed near its origin (**arrows**). These changes, however, may represent arteriosclerosis rather than tumor encasement. The changes within the anterior pancreaticoduodenal artery are consistent with adenocarcinoma.

(*Continued.*)

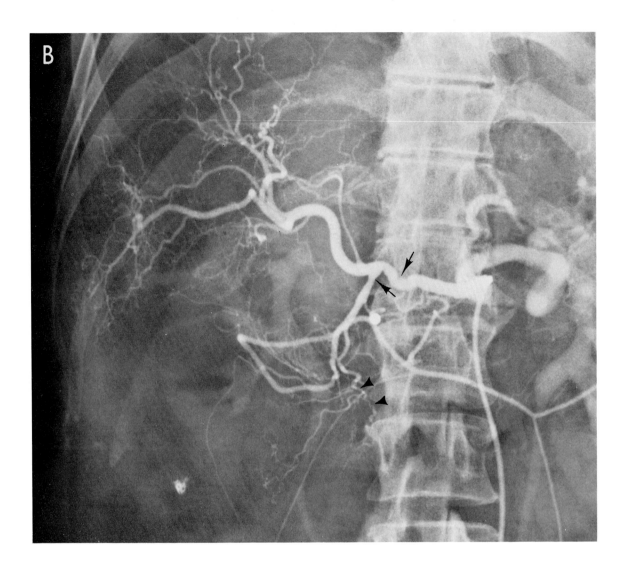

Figure 6 · Adenocarcinoma / 31

Figure 6 (cont.).—Adenocarcinoma of the pancreas.

C, selective celiac angiogram, venous phase: The splenic and portal veins appear to be normal, with numerous lucencies within the portal hepatogram due to avascular metastases (**arrows**).

Comment: Arterial beading, narrowing, tortuosity and obstruction may be found in both adenocarcinoma of the pancreas and pancreatitis. The difficulty in differentiation in this instance is resolved by the finding of an obstructing duodenal lesion and avascular metastases in the liver.

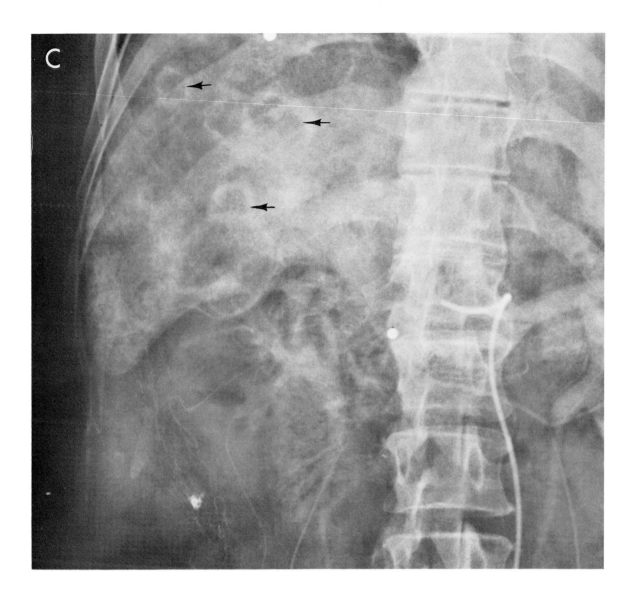

Figure 6 · Adenocarcinoma / 33

Figure 7.—Adenocarcinoma of the pancreas.

A 66-year-old man complained of epigastric pain radiating to the back, a 20-lb weight loss and melena of four-months' duration.

A, right anterior oblique spot film of the duodenum: The mucosal outline of the internal aspect of the duodenal loop is smoothly obliterated (**arrowheads**) by a large mass in the head of the pancreas. A double contour is present inferiorly (**arrows**). The loop is greatly enlarged.

B, selective celiac angiogram, arterial phase: The gastroduodenal artery is irregularly narrowed, completely obstructed distally (**arrowheads**), straightened and displaced laterally. The common hepatic artery (**arrow**) is elevated.

(*Continued.*)

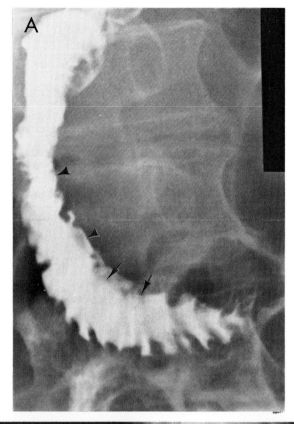
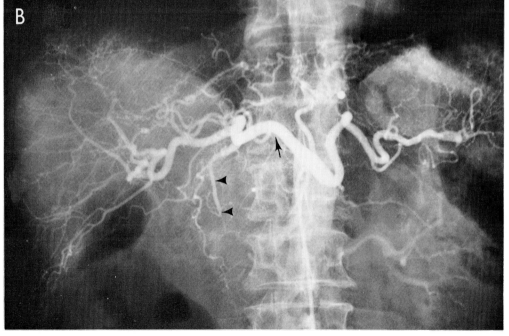

Figure 7 · Adenocarcinoma / 35

Figure 7 (cont.).—Adenocarcinoma of the pancreas.

C, selective celiac angiogram, venous phase: The splenic vein, near the spleen, is completely obstructed (**arrow**). The gallbladder is outlined and is distended.

D, selective superior mesenteric angiogram, arterial phase: Note extensive narrowing of the proximal superior mesenteric artery approximately 3 cm from its origin (**arrowheads**). This has the appearance of tumor encasement rather than arteriosclerosis.

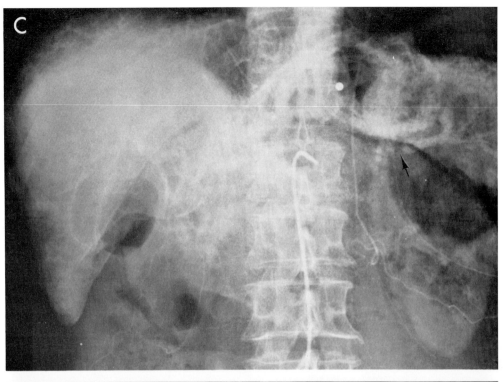
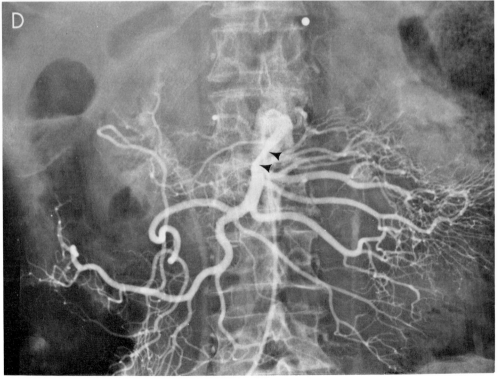

Figure 7 · Adenocarcinoma / 37

Figure 8.—Adenocarcinoma of the pancreas.

Selective celiac angiogram, serial magnification study of the distal pancreatic vasculature in arterial phase: Delineation of the smaller intrahepatic branches is greatly enhanced by the serial magnification technique. Numerous small segments of arterial marginal irregularity and narrowing are demonstrated (**arrows**).

Comment: When large artery encasement and major vein obstruction are angiographically demonstrated in carcinoma of the pancreas, the tumor is in an advanced stage and has widely extended. Such angiographic findings usually indicate a tumor too advanced for surgical resectability. If a higher surgical success rate is to be achieved, the tumor must be diagnosed at an earlier stage. Angiographic methods that enhance small vessel opacification, such as superselective injections and serial magnification studies, may lead to earlier diagnosis at a stage in which only the small arteries are involved. At this stage, however, the arterial changes are difficult to differentiate from those often seen with pancreatitis and arteriosclerosis.

Figure 8, courtesy of Dr. S. Baum, Massachusetts General Hospital, Boston; reprinted from Buranasiri, S., and Baum, S.: Radiology 102:19, January, 1972.

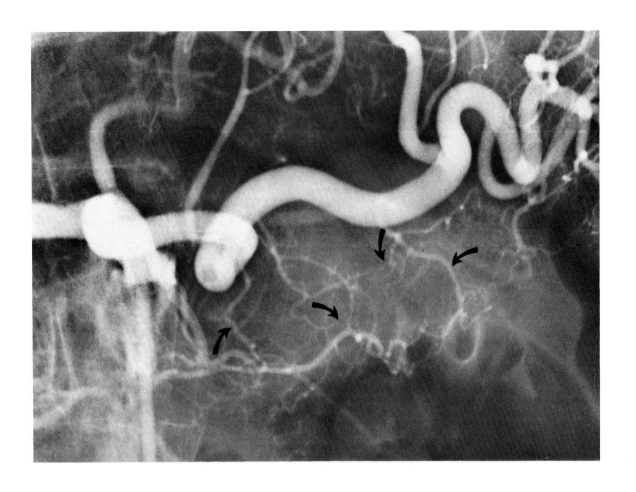

Figure 8 · Adenocarcinoma / 39

Figure 9.—Adenocarcinoma of the pancreas.

A, selective celiac angiogram, arterial phase: The arterial phase of the celiac angiogram appears to be normal. The gastroduodenal artery is not demonstrated and was later proved to be a branch of an accessory right hepatic artery arising from the superior mesenteric artery.

B, selective celiac angiogram, venous phase: The proximal portal vein is narrowed because of extrinsic compression immediately distal to its junction with the splenic vein (**arrows**).

Comment: Involvement of the major vein with no accompanying arterial changes is unusual but does occur. The veins, having thinner walls, are more susceptible to compression and obstruction from an extrinsic mass.

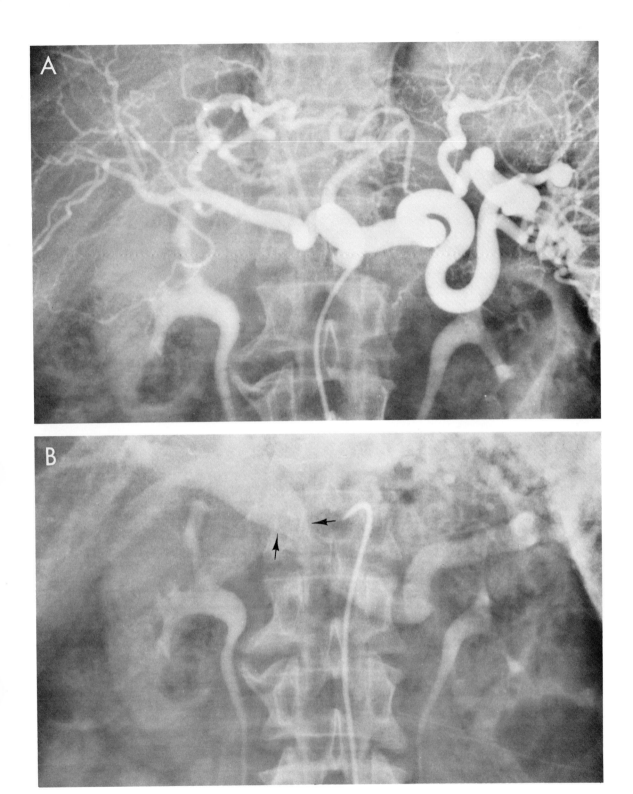

Figure 9 · Adenocarcinoma / 41

Figure 10.—Adenocarcinoma of the pancreas.

A 70-year-old man complained of anorexia, epigastric pain and a 10-lb weight loss over a period of two months. At operation, a large tumor of the tail of the pancreas was found fixed to the posterior abdominal wall.

A, selective celiac angiogram, arterial phase: Note marked narrowing and tortuosity of the distal splenic artery (**arrow**) in the region of the tail of the pancreas. The arterial phase is otherwise normal.

B, selective celiac angiogram, venous phase: The splenic vein is completely obstructed. A large tortuous collateral vein, the gastroepiploic (**arrows**), is evident. The portal vein at this stage is minimally opacified. Some ill-defined lucencies are evident within the hepatogram, indicating metastases.

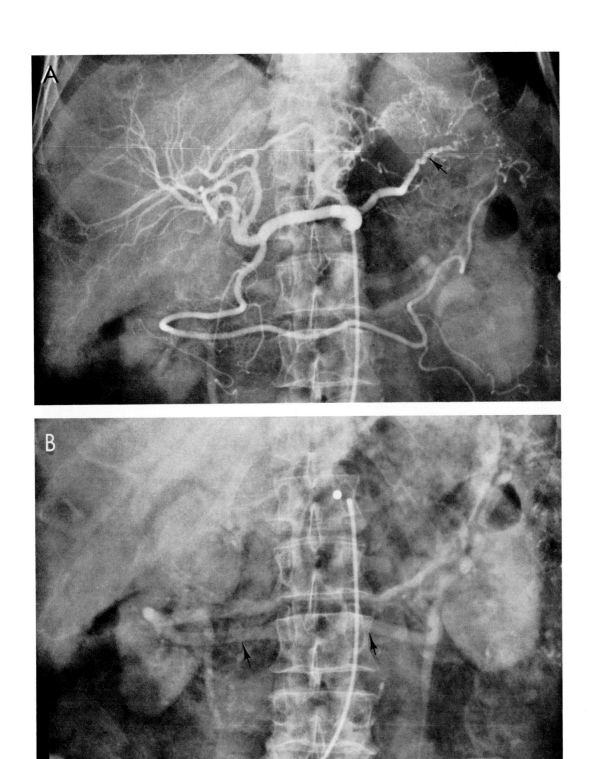

Figure 10 · Adenocarcinoma / 43

Figure 11.—Adenocarcinoma of the pancreas.

A, selective celiac angiogram, arterial phase: Note narrowing and irregularity of the proximal gastroduodenal artery (**arrowheads**). Separation of the cystic artery indicates dilatation of the gallbladder (**arrows**).

B, selective superior mesenteric angiogram, arterial phase: The superior mesenteric artery is narrowed immediately distal to its origin. Poststenotic dilatation is present.

C, selective superior mesenteric angiogram, venous phase: Complete obstruction of the superior mesenteric vein is visible at the level of the pancreas (**arrowheads**). Venous flow to the portal vein is by way of numerous dilated collateral veins that are well opacified (**arrow**). The portal vein and its branches are demonstrated.

Tolazoline hydrochloride (Priscoline), 30 mg, was injected into the superior mesenteric artery before the injection of the opaque medium to enhance the venous opacification.

Figure 11, courtesy of Dr. S. Baum, Massachusetts General Hospital, Boston.

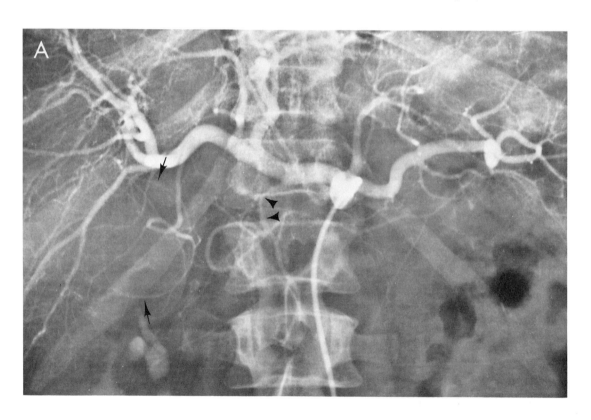

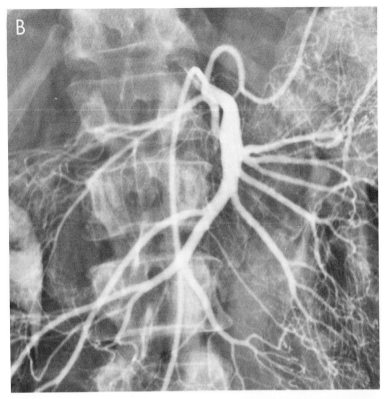

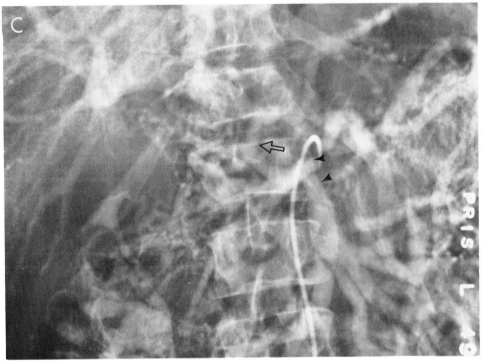

Figure 11 · Adenocarcinoma / 45

Figure 12.—Adenocarcinoma of the pancreas.

A, selective celiac angiogram, arterial phase: The proximal common hepatic artery is narrowed for a distance of 2 cm near its origin (**arrowheads**), representing tumor encasement. The gastroduodenal artery is deviated laterally and is incompletely opacified (**arrow**).

B, selective celiac angiogram, venous phase: The splenic vein is obstructed. Tortuous, dilated collateral veins are opacified. The portal vein is not seen at this stage.

(*Continued.*)

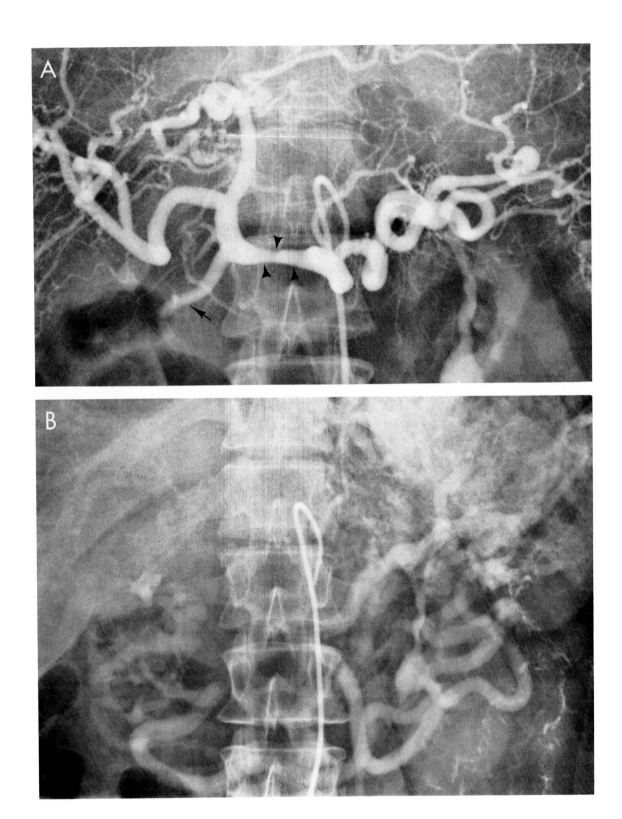

Figure 12 · Adenocarcinoma / 47

Figure 12. (cont.).—Adenocarcinoma of the pancreas.

C, selective superior mesenteric angiogram, arterial phase: The arterial phase is normal.

D, selective superior mesenteric angiogram, venous phase: The superior mesenteric vein is obstructed at the level of the head of the pancreas (**arrowheads**). Venous flow to the portal vein is achieved by large collateral veins (**arrow**), and the portal vein is well opacified.

Comment: This case illustrates the fact that venous changes may be more pronounced than arterial changes in carcinoma of the pancreas. When it is difficult to differentiate arteriosclerosis from the changes of carcinoma of the pancreas in the arterial phase, observation of venous involvement indicates the latter diagnosis.

Figure 12, courtesy of Dr. S. Baum, Massachusetts General Hospital, Boston.

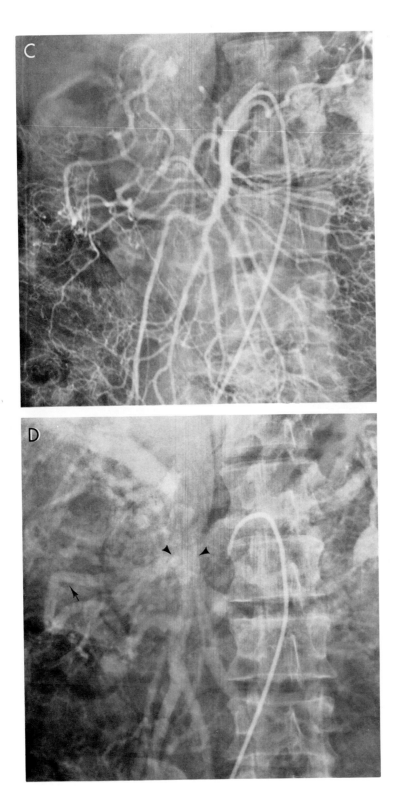

Figure 12 · Adenocarcinoma / 49

Figure 13.—Adenocarcinoma of the pancreas.

A 65-year-old woman had marked weight loss, decreased appetite and severe abdominal pain for several months. Carcinoma involving the head and body of the pancreas was found at operation.

A, selective celiac angiogram, arterial phase: Extensive narrowing of the common hepatic artery at its origin from the celiac axis (**arrow**) indicates tumor encasement.

B, selective superior mesenteric angiogram, arterial phase: There are tortuosity and irregular narrowing of the proximal portions of the arteries arising from the superior mesenteric artery in the region of the pancreas (**arrows**). The superior mesenteric artery shows some irregularity in its inferior wall, indicative of tumor encasement.

(*Continued.*)

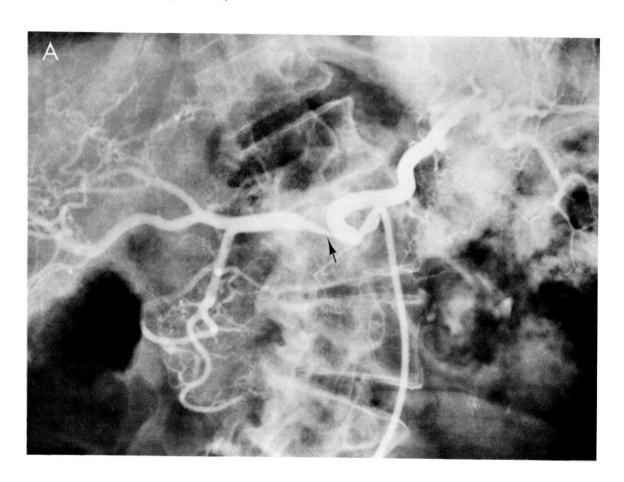

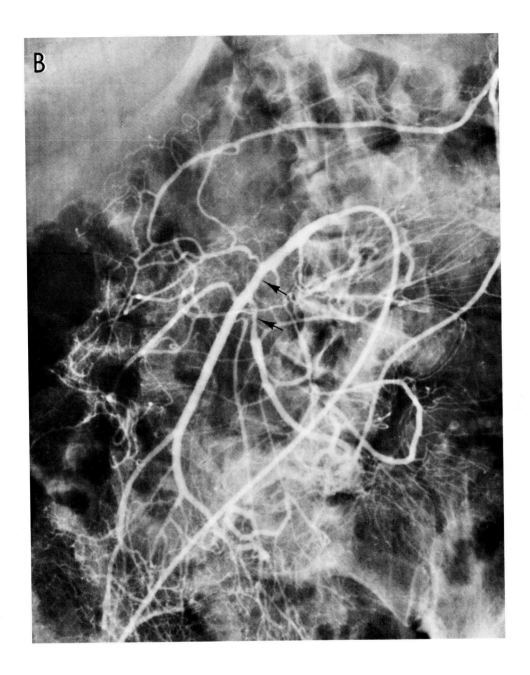

Figure 13 · Adenocarcinoma / 51

Figure 13 (cont.).—Adenocarcinoma of the pancreas.

C, selective superior mesenteric angiogram, venous phase: The superior mesenteric vein is completely obstructed at the level of the pancreas (**arrows**); a collateral vein circumvents the obstruction and opacifies the portal vein.

Figure 13, courtesy of Dr. S. Baum, Massachusetts General Hospital, Boston.

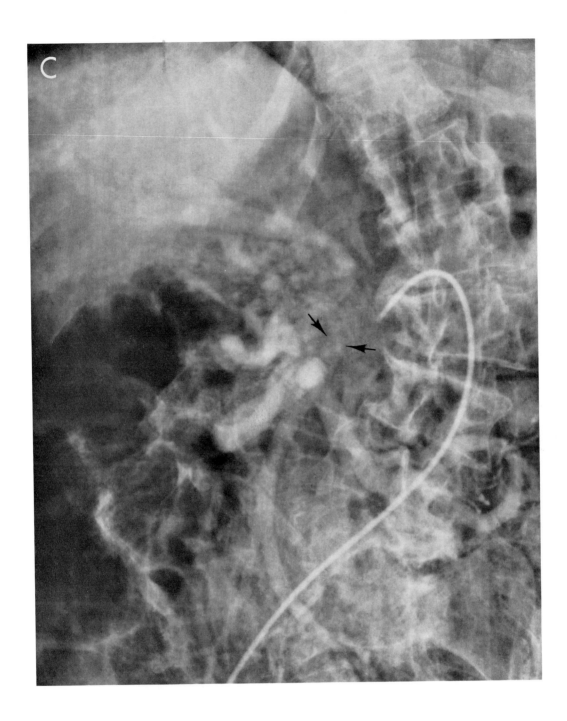

Figure 14.—Adenocarcinoma of the pancreas.

A 73-year-old man had a 45-lb weight loss and anorexia over a six-month period before hospitalization. At operation, carcinoma involving mainly the body of the pancreas was found.

A, selective celiac angiogram, arterial phase: Note the long segment of narrowing of the proximal splenic artery. The artery has a serrated, irregular appearance in this narrowed portion (**arrowheads**); the proximal left gastric artery is also narrowed. The dorsal pancreatic artery is not visible.

B, selective celiac angiogram, venous phase: The splenic vein is completely obstructed at its midportion (**arrowhead**), and the terminal portion is smoothly tapered. Distended collateral veins in the fundus of the stomach are indistinctly delineated.

C, selective superior mesenteric angiogram, venous phase: The superior mesenteric vein is completely obstructed at the level of the pancreas (**arrows**). The portal vein fills by way of the collateral veins which circumvent the obstruction.

Figure 14, courtesy of Dr. S. Baum, Massachusetts General Hospital, Boston.

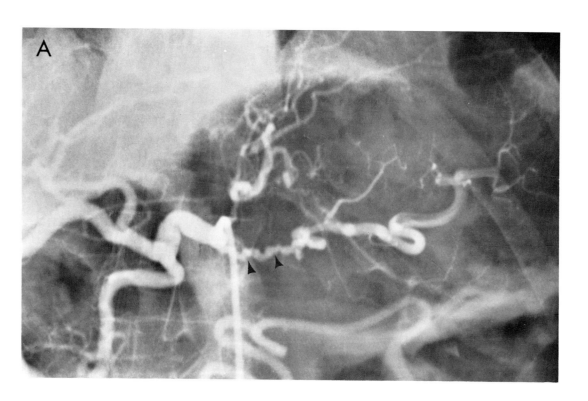

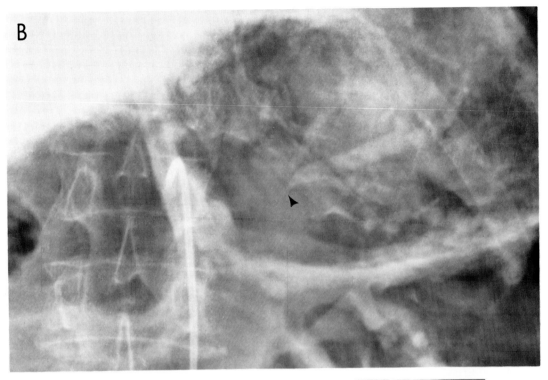
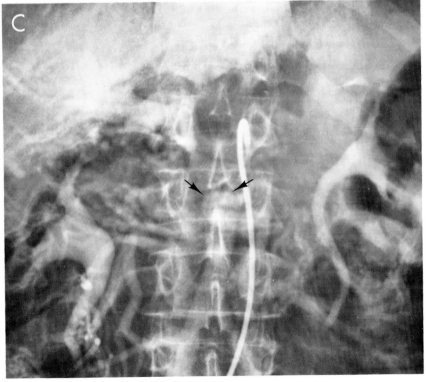

Figure 14 · Adenocarcinoma / 55

Figure 15.—Adenocarcinoma of the pancreas.

A 55-year-old man complained of back pain, radiating subcostally to the epigastric region, of 13 weeks' duration. He was first seen by a neurosurgeon because of the back pain. He had had recent weight loss. Physical examination revealed epigastric guarding and tenderness.

Spot film of the duodenal loop: Note mucosal ulceration in the lateral aspect of the third portion of the duodenum (**arrowheads**). The remaining mucosa in the third portion of the duodenum is abnormal, appearing to be thickened and rigid. Evidence of extrinsic pressure is seen on the medial aspect of the second portion of the duodenum (**arrows**).

Comment: Carcinoma of the pancreas may invade the duodenum, causing mucosal obstruction and occasionally giving the appearance of an intraluminal tumor.

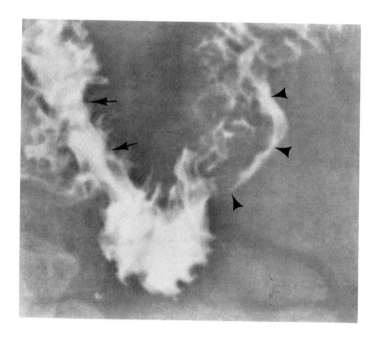

Figure 16.—Adenocarcinoma of the pancreas.

A 58-year-old woman complained of right upper quadrant abdominal and back pain. She had lost 10 lb, and there was a recent onset of jaundice. Physical examination revealed a palpable gallbladder and a separate large epigastric mass. At operation, carcinoma arising in the head of the pancreas was found.

Percutaneous transhepatic cholangiogram, right anterior oblique projection: The common bile duct is completely obstructed at the level of the head of the pancreas. The distal common bile duct tapers to the point of obstruction. Note the smooth tapered segment (**arrowheads**). The entire biliary tree is moderately dilated.

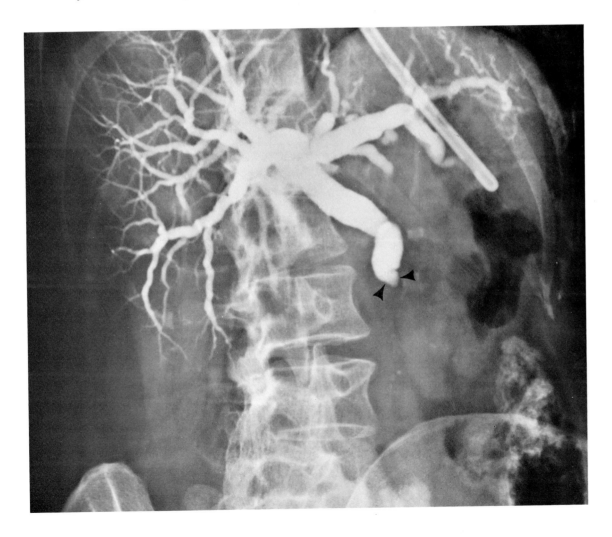

Figure 17.—Adenocarcinoma of the pancreas.

Spot film of the duodenal loop: A segmental, concentric narrowing of the mid-descending duodenum is seen, with a large ulcer on the medial aspect of this segment (**arrowheads**).

Comment: Carcinoma of the head of the pancreas may encircle the descending portion of the duodenum and cause a concentric constriction in this region. In this instance an associated ulceration is caused by invasion of the tumor.

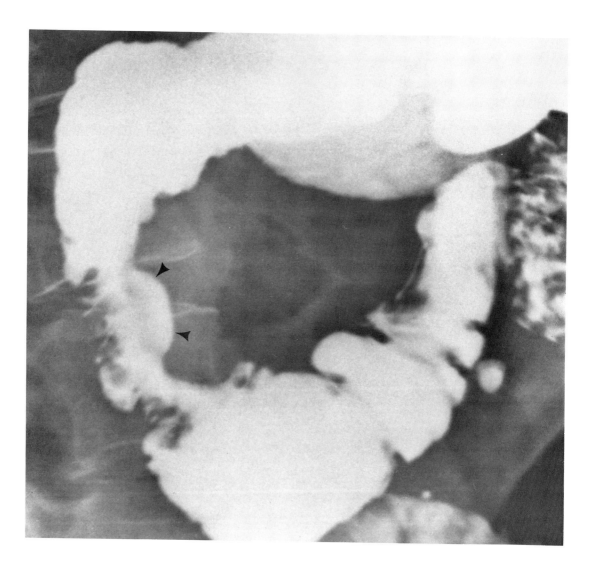

Figure 18.—Adenocarcinoma of the pancreas.

Spot film of stomach and duodenum: The duodenal loop is enlarged. Elevation and extrinsic pressure are present on the distal antrum and the duodenal bulb on the greater curvature aspect (**arrows**). The postbulbar portion of the duodenum is greatly constricted (**arrowheads**), with straightening of the mucosal pattern on the medial aspect of the descending portion.

Comment: Carcinoma of the head of the pancreas was found at operation on this 28 year-old woman.

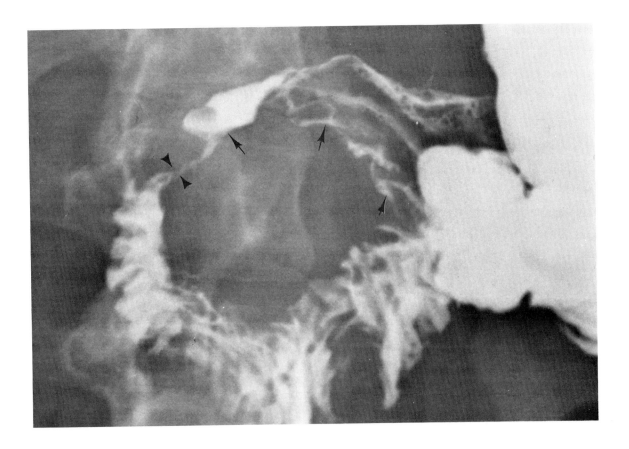

Figure 19.—Adenocarcinoma of the pancreas.

A 62-year-old man had been hospitalized on many occasions for chronic relapsing pancreatitis. Recently, severe epigastric pain and progressive weight loss had recurred.

On physical examination the epigastrium was tender, and the liver was palpable 3 cm below the right costal margin. Operation revealed a carcinoma of the head of the pancreas.

A, intravenous cholangiogram, anteroposterior projection: Extensive calcification is visible throughout the pancreas. The common bile duct is obstructed immediately superior to the pancreatic calcification with a smooth and tapering termination (**arrowheads**) and with dilatation proximal to the obstruction.

B, upper gastrointestinal radiograph, right anterior oblique projection: A well-defined extrinsic pressure defect on the duodenum is present in the postbulbar portion (**arrowheads**), corresponding to the position of the dilated common bile duct. The duodenal mucosa distal to this is nodular in character.

Comment: Although calcification in the pancreas is indicative of inflammatory disease, it does not exclude the presence of neoplasm.

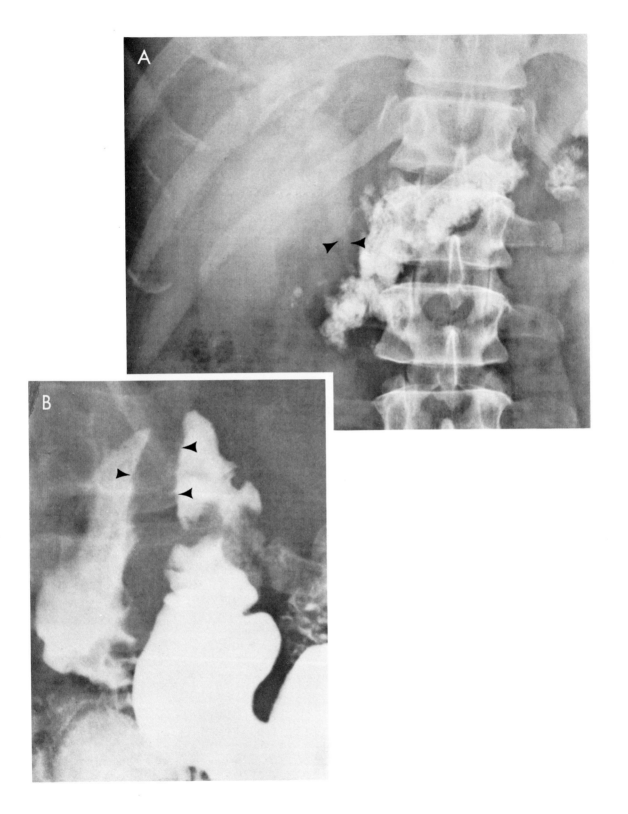

Figure 19 · Adenocarcinoma / 61

Figure 20.—Adenocarcinoma of the head of the pancreas.

A 43-year-old woman complained of epigastric pain penetrating to the back for four months and a 20-lb weight loss. Physical examination was unrevealing. A 7-cm tumor involving the body and tail of the pancreas was found to be unresectable.

Lateral spot film of stomach: A well-demarcated indentation of the posterior wall of the stomach (**arrows**) with irregularity of the mucosa in this region represents infiltration by carcinoma of the pancreas.

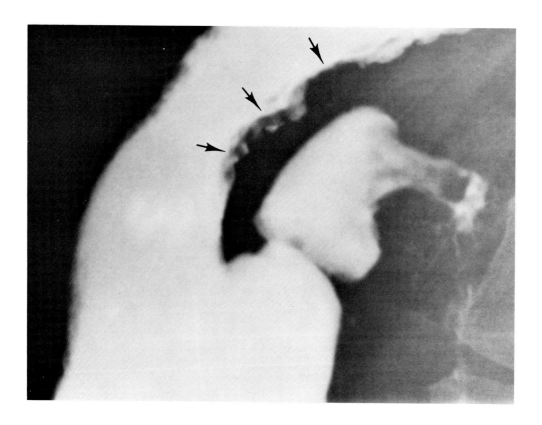

Figure 21.—Adenocarcinoma of the pancreas.

A 68-year-old man had abdominal distress of long duration, with recent exacerbations. The liver was palpable, hard and irregular. Operation revealed adenocarcinoma of the body of the pancreas.

Upper gastrointestinal radiograph, left posterior oblique projection: Note anterior and lateral displacement of the fundus and proximal body of the stomach with evidence of extrinsic pressure on the stomach in this region (**arrows**).

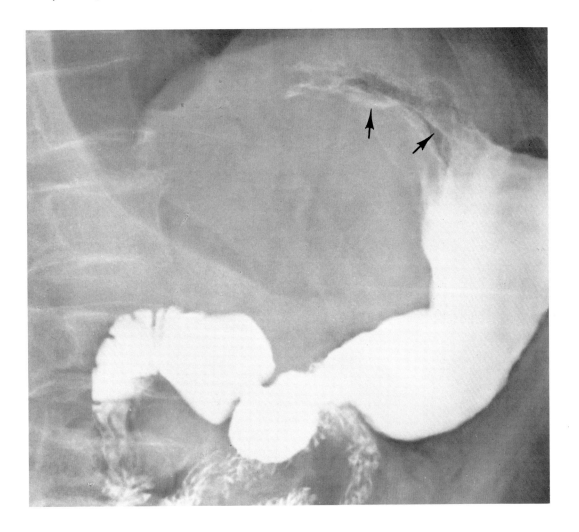

Figure 22.—Adenocarcinoma of the pancreas.

A 58-year-old man had a 13-week history of jaundice, abdominal pain, anorexia, 30-lb weight loss and recent onset of diabetes. At operation, a large adenocarcinoma was found involving the head of the pancreas.

Upper gastrointestinal radiograph, right anterior oblique spot film: There is straightening of the mucosal border on the medial aspect of the descending portion of the duodenum (**arrowheads**). A double contour in this region (**arrows**) is the result of a mass in the head of the pancreas pressing on the loop. The gastric antrum is elevated.

Figure 22 · Adenocarcinoma / 65

Figure 23.—Adenocarcinoma of the pancreas.

A 61-year-old woman had upper abdominal discomfort plus pain in the back and shoulders of three-months' duration. She had accompanying anorexia, nausea and a 10-lb weight loss. Physical examination was not revealing. At operation a large adenocarcinoma was found involving the entire pancreas and extending to the hiatal opening with invasion of the distal esophagus and the stomach.

Spot film of esophagogastric junction: There is distortion of the distal esophagus and the fundus of the stomach at the cardia. The fundus of the stomach is displaced laterally (**arrows**). The mucosa in the distal esophagus appears to be nodular and partially destroyed (**arrowheads**).

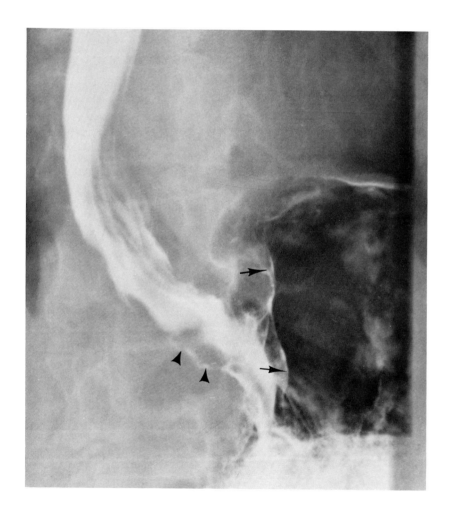

Figure 24.—Adenocarcinoma of the pancreas.

A 48-year-old woman was hospitalized for investigation of a chronic cough and weakness. An adenocarcinoma of the head of the pancreas was found at operation.

Upper gastrointestinal radiograph, anteroposterior projection: A large filling defect is present in the second portion of the duodenum (**arrowheads**), having the appearance of an extrinsic nodular mass stretching and separating the overlying intact mucosa.

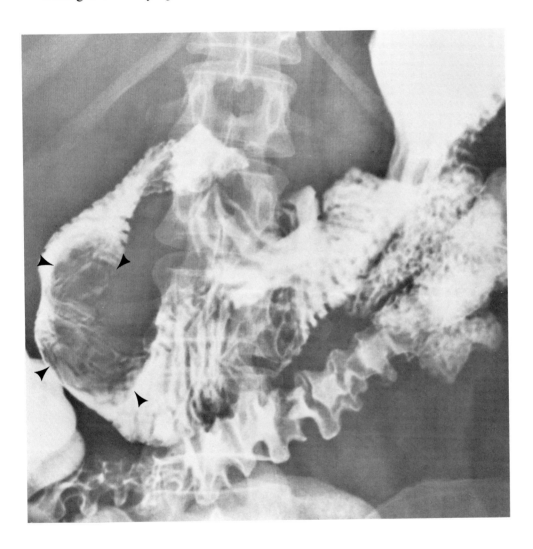

Figure 25.—Adenocarcinoma of the pancreas.

A 73-year-old man had polyuria, thirst and weight loss for two months. Onset of jaundice was noted three weeks before hospitalization.

On physical examination the gallbladder was palpably enlarged, and hepatomegaly was also noted. The adenocarcinoma involved the body and head of the pancreas.

Upper gastrointestinal radiograph, right anterior oblique spot film: The mucosa of the duodenal loop is thickened, rigid and separated in the descending portion (**arrowheads**). A reverse-figure-3 sign of the medial aspect of the descending duodenum is present as a result of extrinsic pressure by a mass in the head of the pancreas.

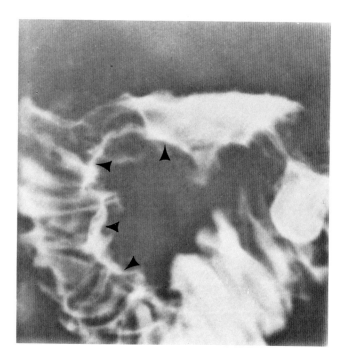

Figure 26.—Adenocarcinoma of the pancreas.

A 51-year-old man was hospitalized seven weeks after a cholecystojejunostomy with increasing upper abdominal pain and jaundice. Physical examination revealed a healed right paramedial scar, jaundice and right epigastric tenderness.

At operation an adenocarcinoma was found in the head of the pancreas and an abscess was found posterior to the pancreatic head. Metastasis was present in the liver and in the celiac axis lymph nodes.

Upper gastrointestinal radiograph, anteroposterior projection: There is elevation of the greater curvature aspect of the distal antrum of the stomach (**arrows**). The distal descending portion of the duodenum is narrowed with extrinsic pressure on the internal aspect (**arrowheads**).

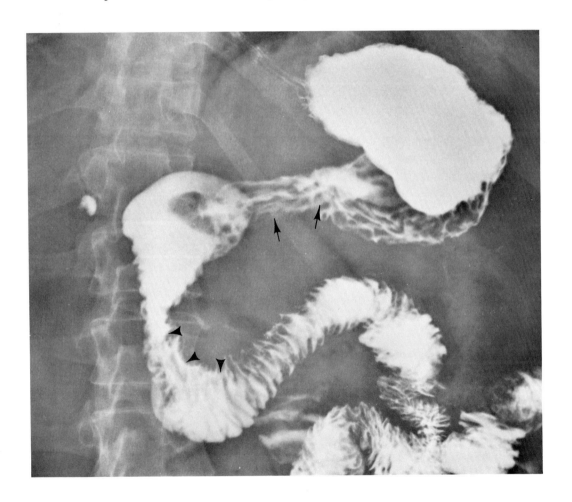

Figure 27.—Adenocarcinoma of the pancreas.

A 68-year-old man had recurrent jaundice six months after a cholecystojejunostomy. At operation a tumor was found to involve the head of the pancreas, the distal common bile duct and the duodenum. Liver metastases were also found.

Upper gastrointestinal tract, right anterior oblique spot film: Note extensive concentric narrowing of the mid-descending portion of the duodenum and extrinsic pressure on the internal aspect of the loop in this region. There appears to be invasion of the mucosa in this segment (**arrows**).

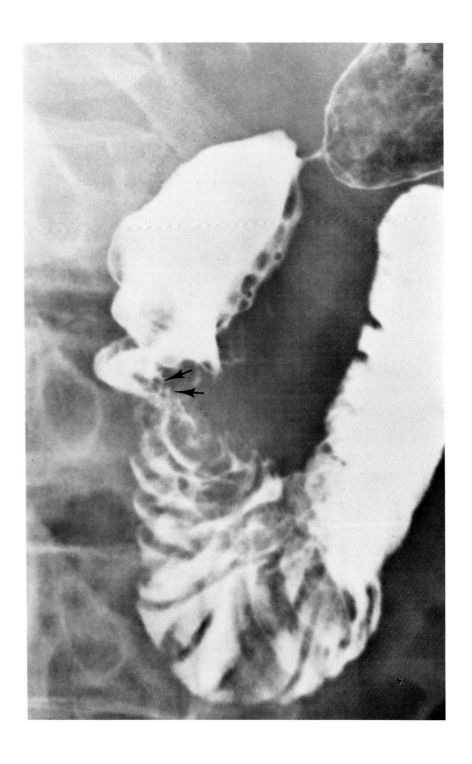

Figure 27 · Adenocarcinoma / 71

Figure 28.—Adenocarcinoma of the pancreas.

A 69-year-old man had anorexia and intractable abdominal pain. Adenocarcinoma of the body and tail of the pancreas was found on surgical exploration.

Upper gastrointestinal radiograph, left posterior oblique projection: The body of the stomach is displaced laterally and anteriorly. An extrinsic mass has produced a pressure defect on the lesser curvature of the stomach, and a large ulcer crater is visible in the superior portion of this segment (**arrowheads**). This is indicative of invasion of the stomach wall by the tumor.

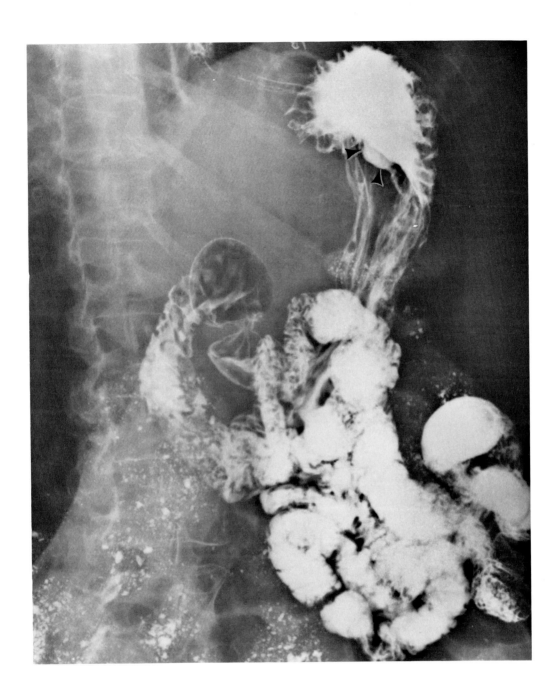

Figure 28 · Adenocarcinoma / 73

Figure 29.—Adenocarcinoma of the pancreas.

A 65-year-old woman had recent onset of epigastric pain and chronic diarrhea. An epigastric mass was palpated on physical examination. At surgical exploration, the pancreas was enlarged, lobulated and cystic, and an adenocarcinoma was found in the head of the pancreas. Metastases were present in the liver.

A, upper gastrointestinal radiograph, anteroposterior projection: There is elevation of the proximal duodenum and pressure on the internal aspect of the duodenal loop (**arrows**) with concentric narrowing in the postbulbar region associated with hypertrophic mucosa.

B, operative cholangiogram, anteroposterior projection: An indentation of the mid-common bile duct, which has the appearance of extrinsic pressure (**arrows**), is apparent, with an associated narrowing causing partial obstruction. The common bile duct and common hepatic duct proximal to the constriction are minimally dilated.

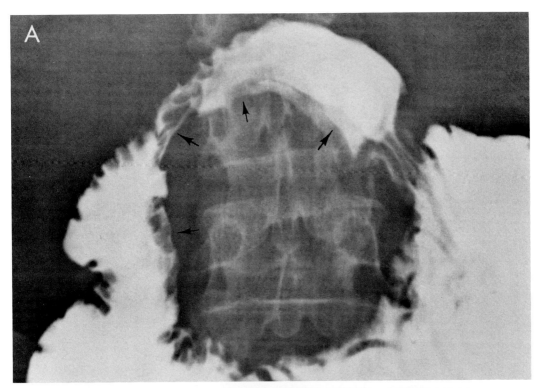

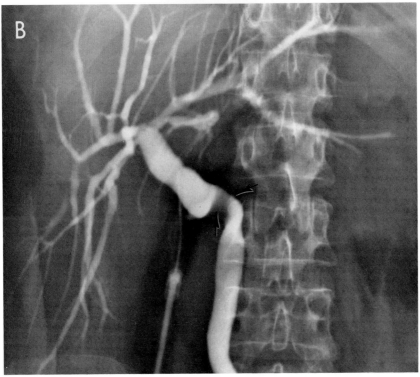

Figure 29 · Adenocarcinoma / 75

Figure 30.—Adenocarcinoma of the pancreas.

T-tube cholangiogram, anteroposterior projection: The distal common bile duct is obstructed, and the biliary tree is dilated proximal to this. The distal common bile duct is smoothly tapered to the point of obstruction (**arrow**).

Comment: This smooth tapering is characteristically caused by carcinoma of the head of the pancreas.

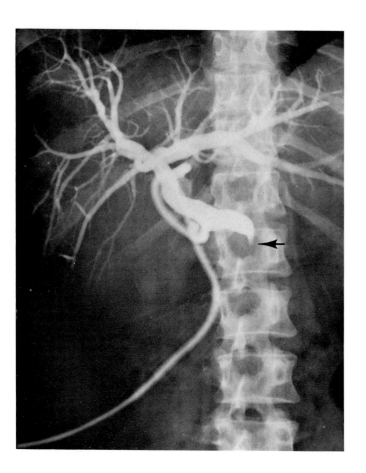

Figure 31.—Normal pancreas.

Selenomethionine scan of the pancreas, anteroposterior projection: The pancreatic outline is complete. Uptake within the liver is also visualized.

Figures 31–37, courtesy of Dr. Antonio Rodriquez-Antunez, Cleveland Clinic, Cleveland.

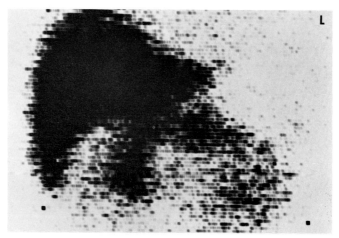

Figure 32.—Carcinoma of the head of the pancreas.

Selenomethionine scan of the pancreas, anteroposterior projection: The outline of the pancreas is incomplete, with total absence of uptake in the head.

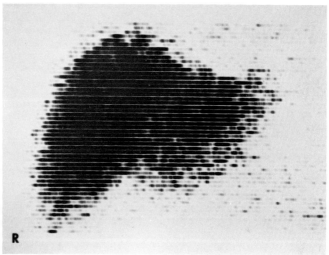

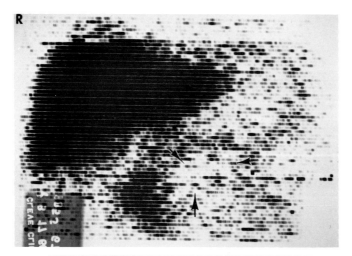

Figure 33.—Adenocarcinoma of the body of the pancreas.

Selenomethionine scan of the pancreas, anteroposterior projection: Note large area of decreased uptake in the midbody of the pancreas (**arrows**).

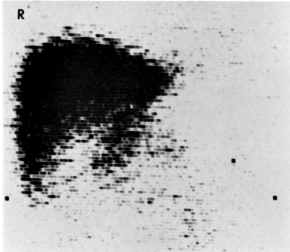

Figure 34.—Adenocarcinoma of the body of the pancreas.

Selenomethionine scan of the pancreas, anteroposterior projection: Note decreased uptake in the distal body of the pancreas.

Figure 35.—Adenocarcinoma of the body of the pancreas.

Selenomethionine scan of the pancreas, anteroposterior projection: There is deficient uptake in the midbody of the pancreas (**arrows**).

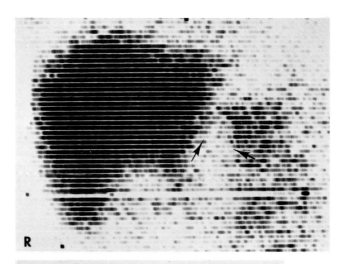

Figure 36.—Adenocarcinoma of the body and tail of the pancreas.

Selenomethionine scan of the pancreas, anteroposterior projection: Note absence of uptake in the distal body and tail of the pancreas (**arrows**).

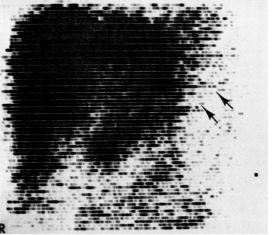

Figure 37.—Adenocarcinoma of the body of the pancreas.

Selenomethionine scan of the pancreas, anteroposterior projection: Note deficient uptake in the body of the pancreas (**arrows**).

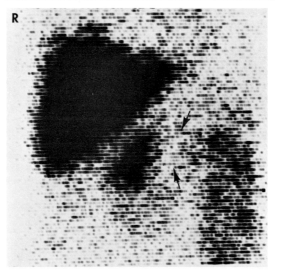

Figure 38.—Cystadenoma of the pancreas.

A 63-year-old woman complained of increasing fullness in the left abdomen for 1½ years. A mass was palpated in the left upper quadrant and a bruit was auscultated over the mass. The patient had a diabetic type of glucose tolerance test.

A, intravenous pyelogram, anteroposterior projection: The left kidney is displaced superiorly by a large soft tissue mass that is roughly circular (**arrows**) and has calcification within. The left psoas outline is obliterated, and the stomach gas bubble is displaced by the mass (**arrowheads**). Displacement of the kidney plus obliteration of the psoas outline indicates that this is a retroperitoneal mass.

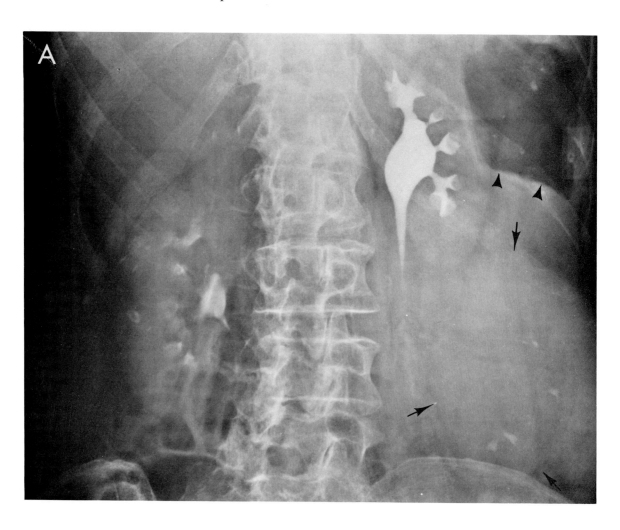

B, upper gastrointestinal radiograph, anteroposterior projection: The distal body and antrum of the stomach and the proximal duodenum are displaced laterally and inferiorly by a large mass (**arrows**) that presses on the lesser curvature aspect.

(*Continued.*)

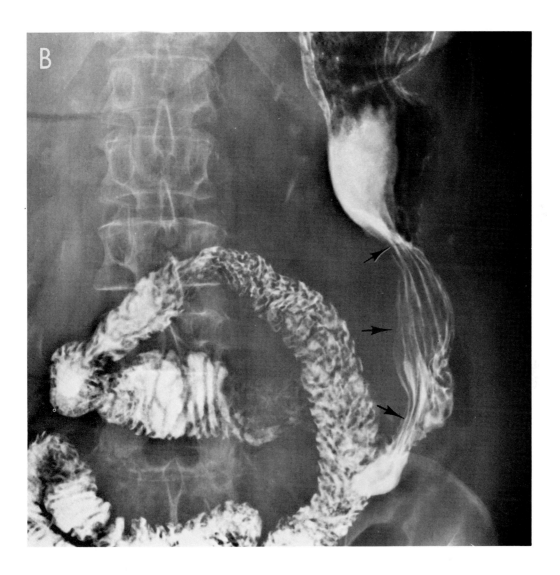

Figure 38 · Cystadenoma / 81

Figure 38 (cont.).—Cystadenoma of the pancreas.

C, selective celiac angiogram, arterial phase: A large branch of the celiac axis, the gastroduodenal artery, descends vertically to the left of the midline. It is greatly enlarged. The dorsal pancreatic artery is also delineated (**arrows**). Both of these arteries give branches to the large mass in the left portion of the abdomen. The vessels within the tumor are abnormal and have the appearance of tumor vascularity.

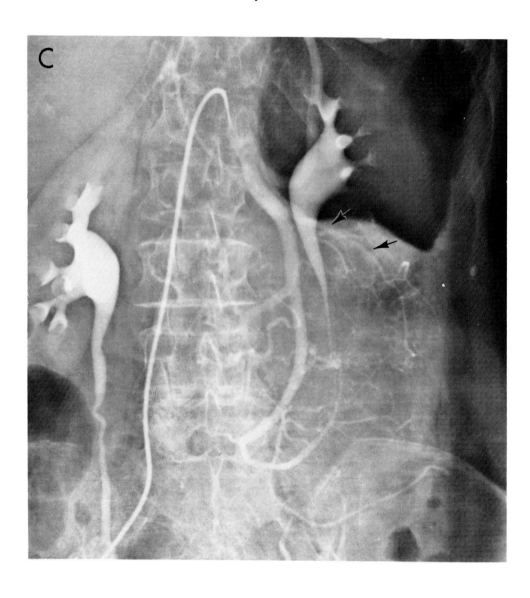

D, selective celiac angiogram, venous phase: The tumor within the left portion of the abdomen shows a dense vascular blush. The splenic vein is obstructed (**arrows**). Large tortuous collateral veins are evident lateral and inferior to the mass (**arrowheads**). The portal vein is opacified by way of the collateral flow.

(*Continued.*)

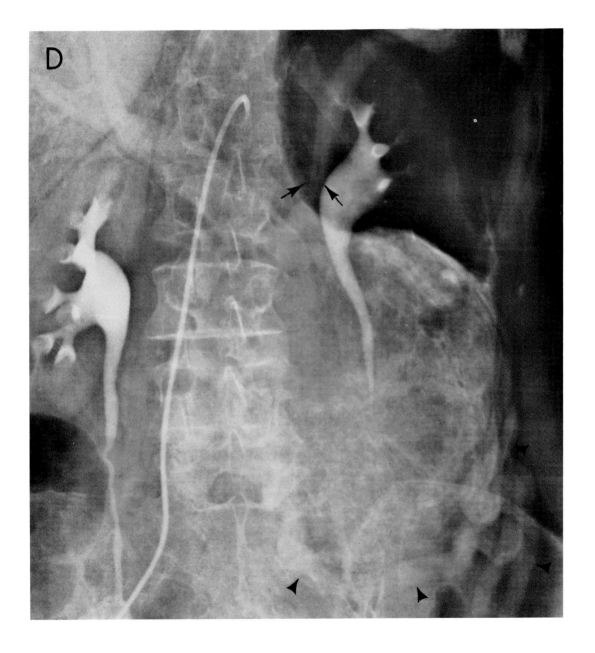

Figure 38 (cont.).—Cystadenoma of the pancreas.

E, selective superior mesenteric angiogram, arterial phase: The right hepatic artery arises from the superior mesenteric artery. The inferior pancreaticoduodenal artery is opacified. This also supplies the extremely vascular pancreatic mass.

Comment: Calcification occurs in 10% of patients with cystadenoma or cystadenocarcinoma of the pancreas. The pattern of calcification may be nonspecific or may have a radiating pattern. Only islet cell tumors of the pancreas are similarly richly vascular. Islet cell tumors are commonly smaller, do not have calcification and may be multiple.

Figure 38, courtesy of Dr. S. Baum, Massachusetts General Hospital, Boston.

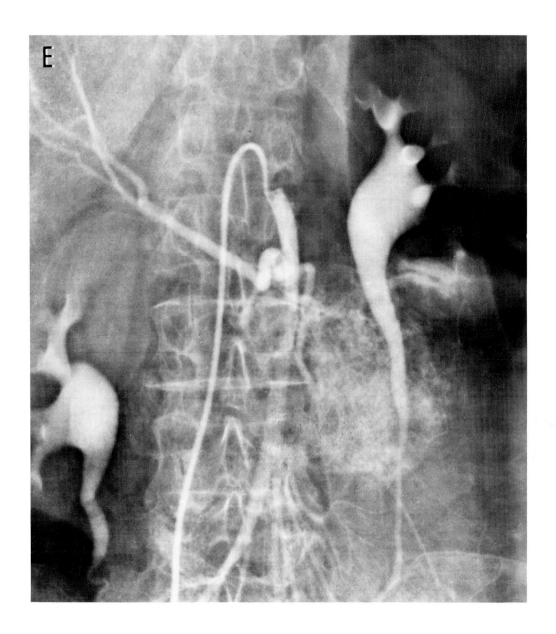

Figure 38 · Cystadenoma / 85

Figure 39.—Islet cell tumor of the pancreas.

A, selective celiac angiogram, arterial phase: Pancreatic branches of the splenic artery supply a richly vascular mass within the pancreas (**arrows**). This appearance suggests tumor vasculature.

B, selective celiac angiogram, venous phase: A tumor blush is still apparent in this late phase (**arrows**). The splenic vein is completely obstructed. Large tortuous collateral veins circumvent the obstruction, and the portal vein is well opacified.

Comment: This islet cell tumor is unusual because of its large size. However, the rich tumor vascularity and dense blush are typical of such tumors.

Figure 39, courtesy of Dr. James J. Pollard, Charles Choate Memorial Hospital, Woburn, Mass.

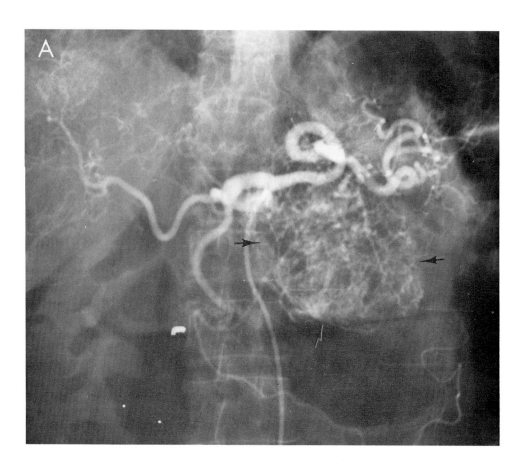

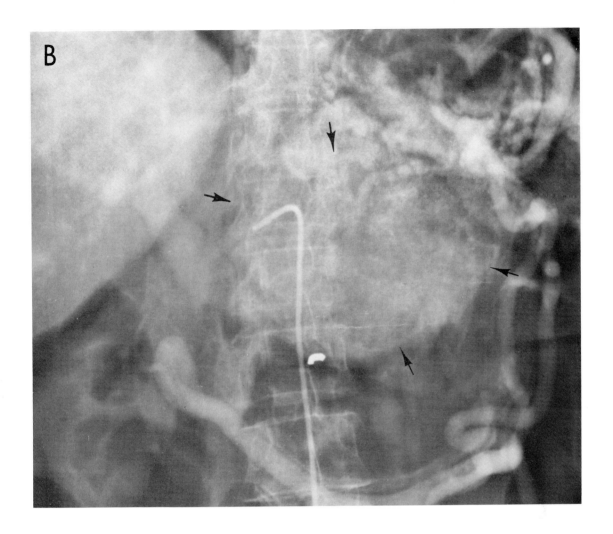

Figure 39 · Islet Cell Tumor / 87

Figure 40.—Mixed islet cell and ductal cell carcinoma of the pancreas.

A 58-year-old woman had many dramatic hypoglycemic episodes.

A, selective celiac angiogram, arterial phase: The caudal pancreatic branches of the splenic artery opacify a small circular lesion within the pancreas (**arrows**) which, although not richly vascular, does show a tumor blush despite the overlying spleen.

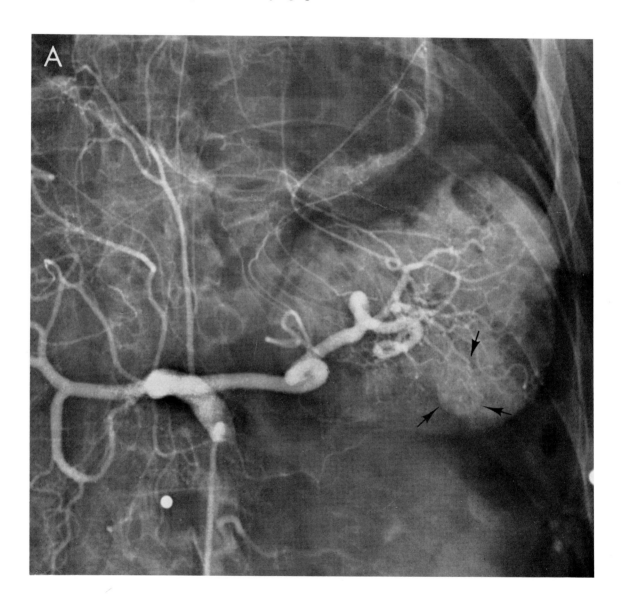

B, selective celiac angiogram, venous phase: The small circular tumor (**arrows**) within the tail of the pancreas is densely opacified as a result of the tumor blush.

Comment: This tumor had both islet cell and ductal cell carcinoma components. Small, circular dense opacities demonstrated during the celiac angiogram may represent evidence of tumor, but this must be differentiated from splenules or lymph nodes which may have a similar appearance.

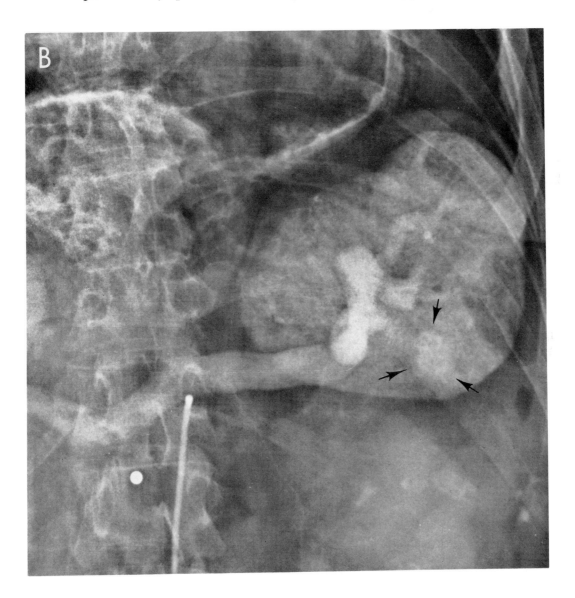

Figure 40 · Mixed Islet Cell & Ductal Tumor

Figure 41.—Benign insulinoma.

A 38-year-old woman had a history of recurring episodes of low blood glucose levels for six years.

A, selective celiac angiogram, arterial phase: Note the small oval density high in the left upper quadrant (**arrows**) overlying the spleen and being supplied by a prominent branch of the splenic artery.

B, selective celiac angiogram, venous phase: The oval density in the left upper quadrant (**arrows**) of the abdomen is more opacified than during the arterial phase. Small tortuous veins, which communicate with the splenic vein, are seen to drain the tumor.

Figure 41, courtesy of Dr. S. Baum, Massachusetts General Hospital, Boston.

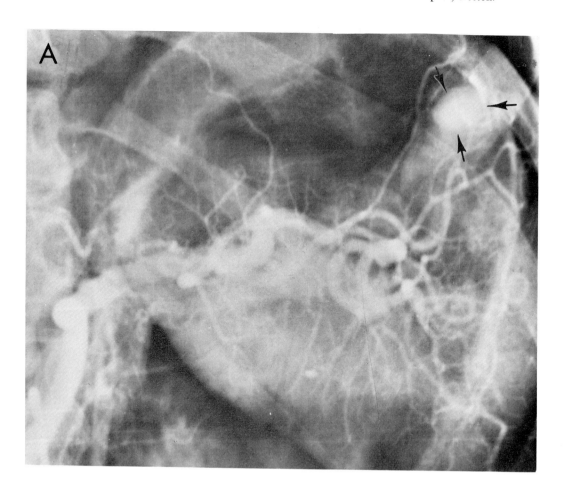

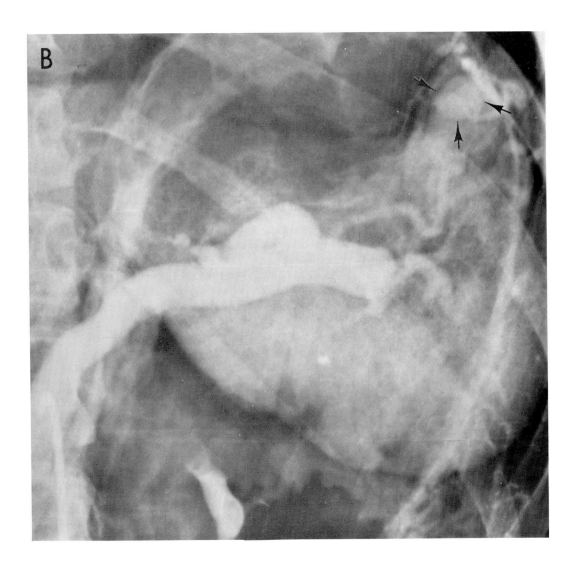

Figure 41 · Benign Insulinoma / 91

Figure 42.—Nonfunctioning islet cell carcinoma of the pancreas.

A 69-year-old man complained of nocturia only. An upper abdominal mass, hepatomegaly and splenomegaly were palpated.

A, upper gastrointestinal radiograph, lateral spot film: Stomach is displaced anteriorly, and a smooth, extrinsic pressure defect on the posterior wall (**arrows**) is caused by a mass impinging on the stomach in this position. Some irregularity of the mucosa is seen in the superior portion of this impression, suggesting invasion of the stomach wall by the tumor.

(*Continued.*)

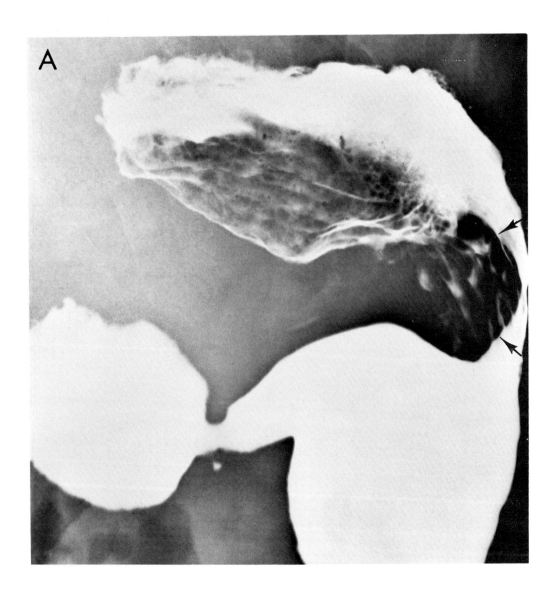

Figure 42 · Nonfunctioning Islet Cell Tumor / 93

Figure 42 (cont.).—Nonfunctioning islet cell carcinoma of the pancreas.

B, selective celiac angiogram, late arterial phase: There is a large vascular mass in the left upper abdomen that receives its blood supply from the pancreatic branches of the splenic artery. The vascular pattern is highly irregular; the appearance is that of tumor vasculature. Small hypervascular nodules, representing metastatic foci, are opacified within the liver **(arrows)**.

C, selective celiac angiogram, venous phase: The tumor and the distal pancreas remain opacified with a dense tumor blush **(arrows)**. The splenic

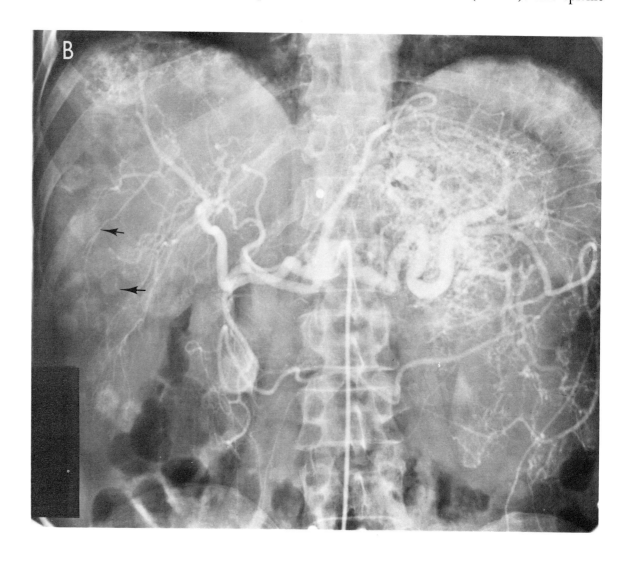

vein is obstructed. Collateral veins circumvent the obstruction (**arrowheads**), and the portal vein is opacified. The late hepatogram is mottled because of metastases within the liver.

Comment: It is usually not possible to determine the character of islet cell tumor of the pancreas from the appearance of the primary lesion. However, when metastases are demonstrated, the diagnosis can be made. Islet cell carcinomas are believed to metastasize to the liver early in their life span. Hepatic metastasis progresses slowly.

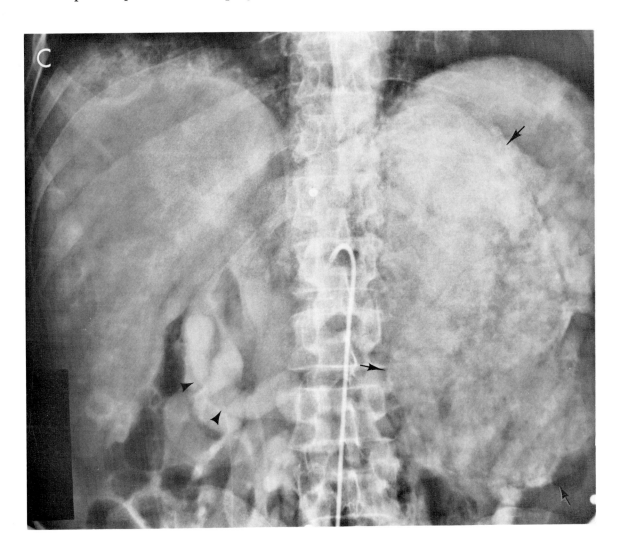

Figure 43.—Insulinoma.

A 36-year-old woman had recurring episodes of double vision, dysarthria, unsteady gait and a feeling of unreality for three years. Physical examination gave negative results. Fasting blood sugar tests and albuminoid tests suggested the possibility of an insulinoma.

A, selective celiac angiogram, arterial phase: Some small irregular vessels are noted distally in the pancreas (**arrowheads**), but dense tumor vascularity is not seen.

B, selective celiac angiogram, venous phase: A dense tumor blush is visible inferior to the midportion of the splenic vein (**arrowheads**).

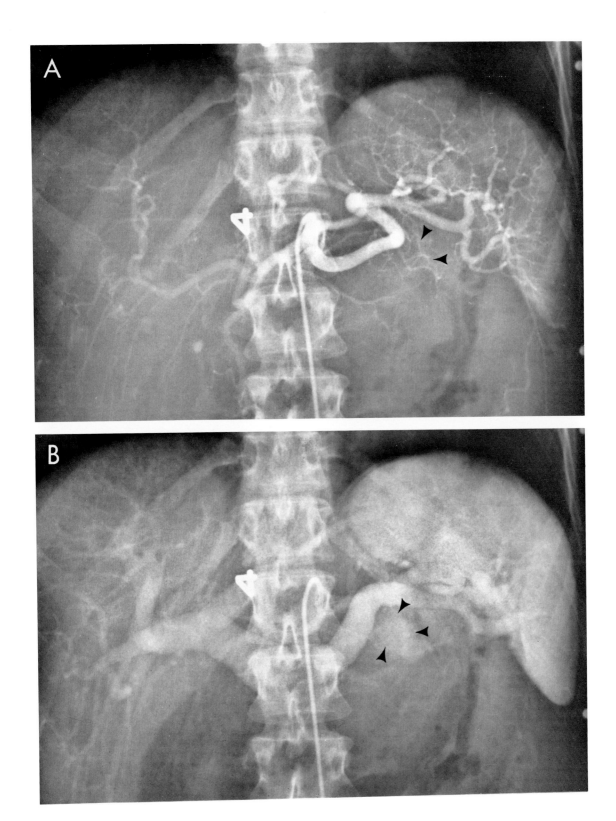

Figure 43 · Insulinoma

Figure 44.—Pseudocyst of the pancreas.

A 32-year-old woman had suffered abdominal trauma in an automobile accident two months previously. She complained of upper abdominal pain and weight loss. A large epigastric mass was palpable.

A, upper gastrointestinal radiograph, anteroposterior projection: The body and antrum of the stomach are displaced laterally and inferiorly by a large mass that causes the smooth, extrinsic impression on the lesser curvature aspect (**arrowheads**).

(*Continued.*)

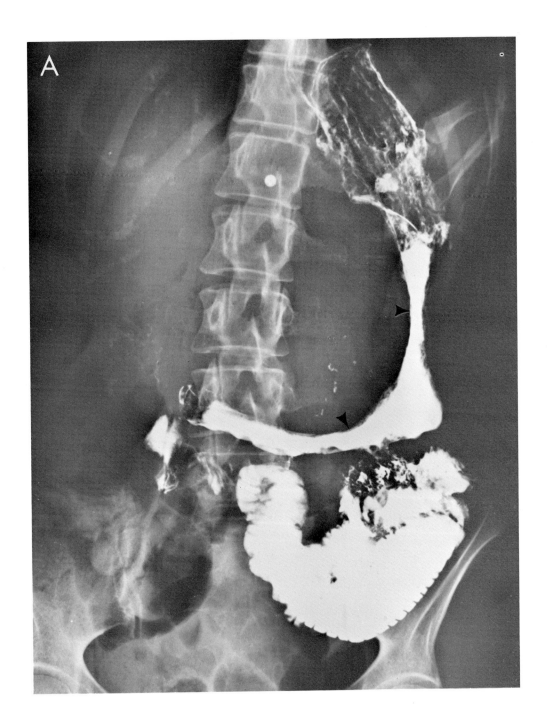

Figure 44 · Pseudocyst / 99

Figure 44 (cont.).—Pseudocyst of the pancreas.

B, selective celiac angiogram, arterial phase: The splenic artery is displaced superiorly. A large artery supplying the stomach on its lesser curvature aspect is smoothly draped over the large mass in the left abdomen (**arrowheads**). The stomach gas shadow is displaced laterally and inferiorly.

C, selective celiac angiogram, venous phase: The mass in the left abdomen is completely avascular. Its displacement of the stomach is again evident. The splenic vein is elevated.

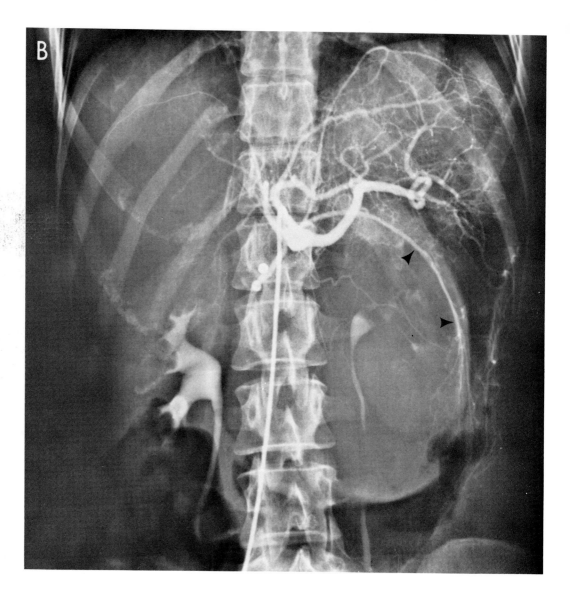

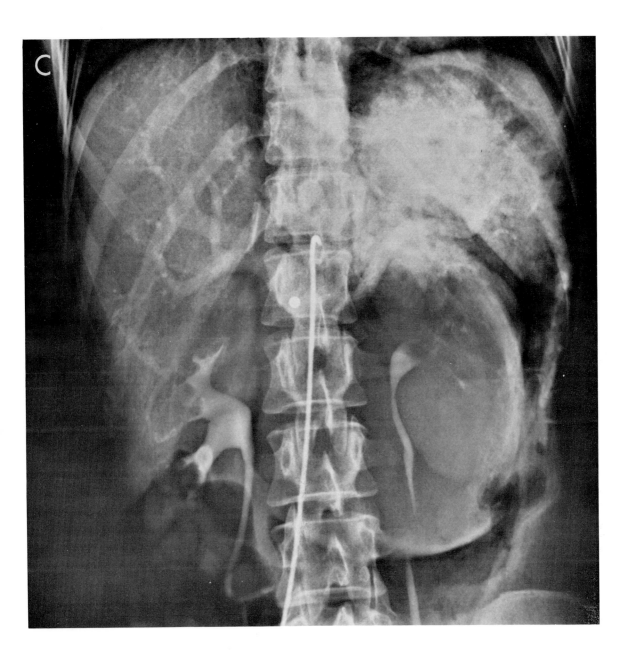

Figure 45.—Pseudocyst of the pancreas.

A 69-year-old man had upper abdominal pain, anorexia and jaundice. He had no history of pancreatitis or abdominal trauma. A large cystic mass was palpable in the epigastrium.

A, upper gastrointestinal radiograph, anteroposterior projection: The body and antrum of the stomach are displaced by a mass in the midabdomen (**arrowheads**).

(*Continued.*)

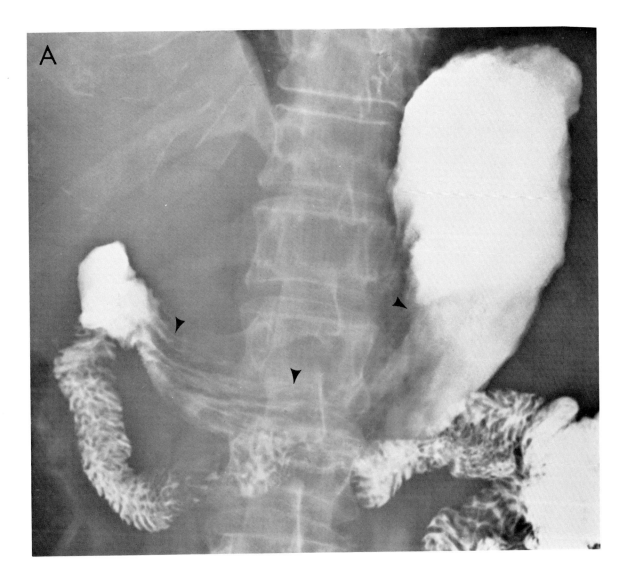

Figure 45 · Pseudocyst

Figure 45 (cont.).—Pseudocyst of the pancreas.

B, selective celiac angiogram, arterial phase: The common hepatic artery is elevated, and the gastroduodenal artery, which appears to be stretched (**arrowheads**), is displaced laterally.

C, selective celiac angiogram, venous phase: The splenic vein is completely obstructed (**arrow**). Collateral veins are present superiorly and inferiorly (**arrowheads**). The portal vein is opacified by way of the collateral veins. The pancreatic mass shows no tumor vascularity or stain.

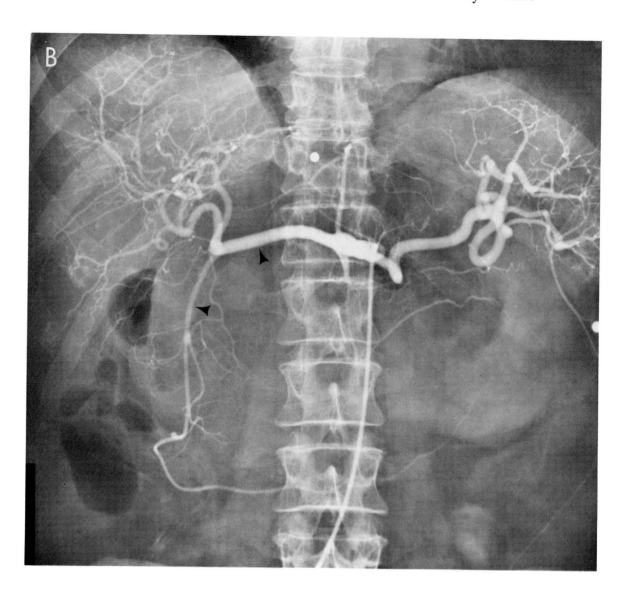

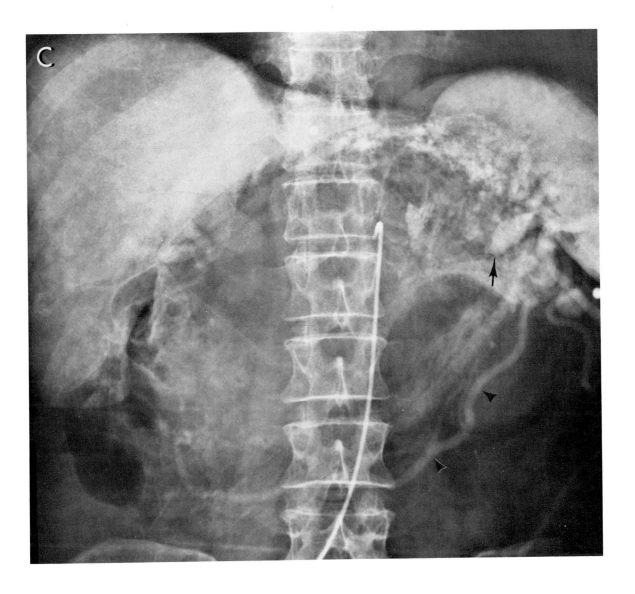

Figure 45 · Pseudocyst

Figure 46.—Cyst of the pancreas.

A 30-year-old woman had had three episodes of acute pancreatitis in the preceding two years. She complained of epigastric pain of recent onset, radiating to the back and increased by eating.

A, upper gastrointestinal radiograph, anteroposterior projection: The duodenal loop is enlarged. Pressure is present on the greater curvature aspect of the gastric antrum and the internal aspect of the duodenal loop (**arrowheads**) from a large mass within the head of the pancreas. Multiple calculi can be seen in the head of the pancreas.

B, operative pancreatogram and cholangiogram, anteroposterior projection: The duct of Wirsung is opacified (**arrowheads**), greatly enlarged, tortuous and serrated in appearance. A collection of contrast material fills a cystic space in the head of the pancreas (**arrows**). The distal common bile duct is not opacified.

C, operative cholangiogram, anteroposterior projection: The gallbladder overlies the proximal common hepatic duct. The distal common bile duct is greatly narrowed and stretched (**arrowheads**). Calcification is visible in the head of the pancreas.

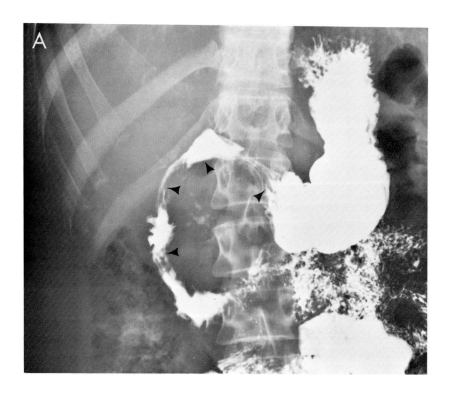

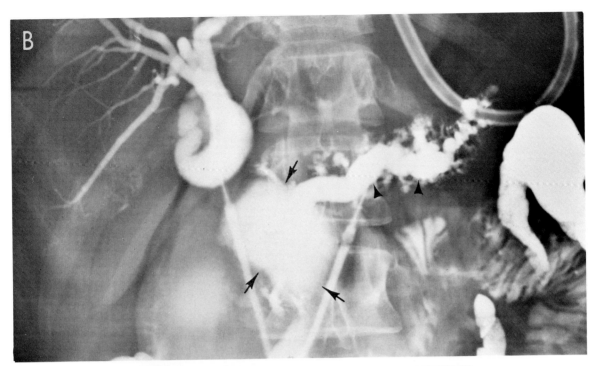

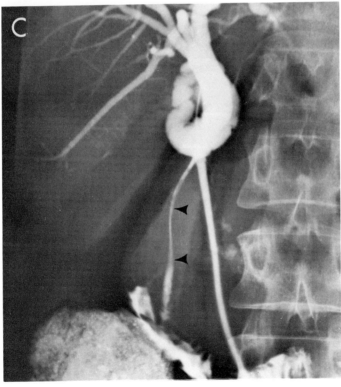

Figure 46 · Cyst / 107

Figure 47.—Pseudocyst of the pancreas.

A 45-year-old woman complained of postprandial epigastric pain, nausea, vomiting, abdominal swelling and anorexia. She had a 15-lb weight loss over a short period. A large pseudocyst was found, situated in the gastrohepatic omentum.

A, upper gastrointestinal radiograph, anteroposterior projection: There is lateral, inferior and anterior displacement of the body and antrum of the stomach by a large pancreatic mass (**arrowheads**). Calcifications are scattered throughout the mass.

B, selective celiac angiogram, arterial phase: The left gastric (**arrowheads**) and splenic arteries are smoothly draped over the large left retroperitoneal mass.

(*Continued.*)

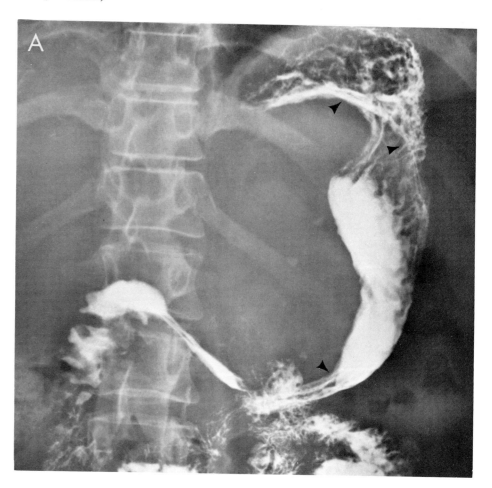

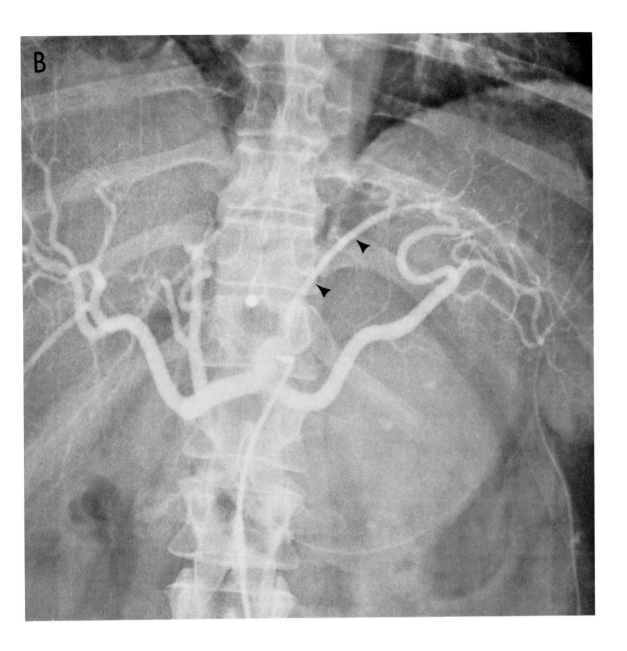

Figure 47 · Pseudocyst / 109

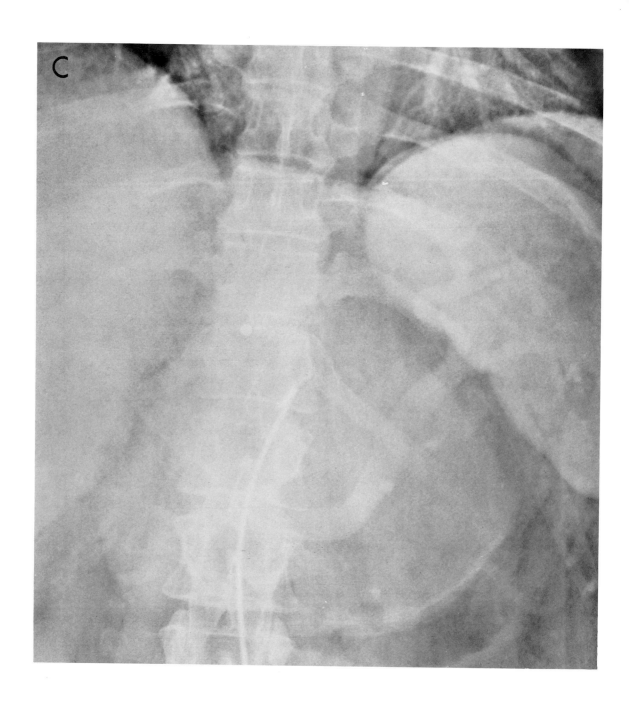

Figure 47 (cont.).—Pseudocyst of the pancreas.

C, selective celiac angiogram, venous phase: The splenic vein is not obstructed. The large pancreatic mass is completely avascular.

D, liver scan, anteroposterior supine projection: Note the area of decreased uptake in the lower margin of the left lobe of the liver, corresponding to extrinsic pressure from the large pancreatic tumor (**arrows**).

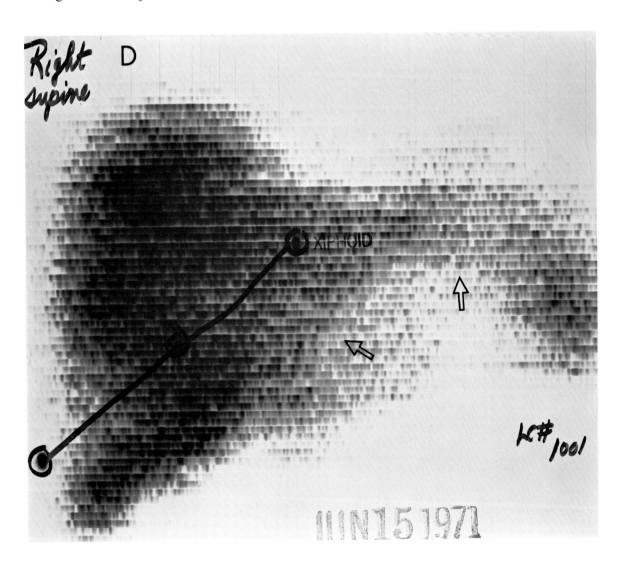

Figure 48.—Pancreatic cyst.

A 76-year-old man with a four-year history of chronic relapsing pancreatitis was seen because of jaundice. At operation an 8-cm cyst was found in the head of the pancreas and adjacent to the common bile duct.

A, operative cholangiogram, anteroposterior projection: The distal common bile duct tapers to a narrowed segment (**arrowheads**) and is partially obstructed at this level. The biliary tract proximal to this is grossly dilated.

B, operative pancreatic cyst injection, anteroposterior projection: Contrast material has been injected into a large cyst in the head of the pancreas during the operation. This cyst was the cause of the narrowing and partial obstruction of the distal common bile duct.

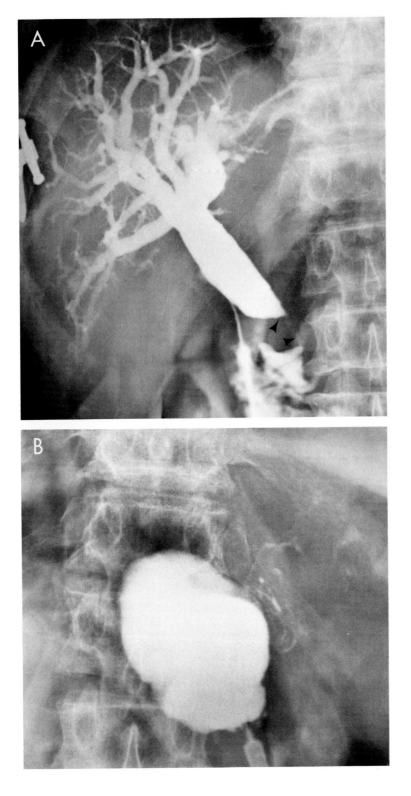

Figure 48 · Cyst / 113

Figure 49.—Metastasis to the pancreas and mesentery from renal cell carcinoma.

A 58-year-old woman had a right nephrectomy for renal cell carcinoma six years before this study. Her present symptoms were weakness, pallor and melena. At laparotomy, a metastatic nodule was found in the greater omentum near the transverse colon, and a second nodule in the small bowel mesentery. The pancreas was completely involved with metastasis, and invasion of the neighboring duodenum was found.

A, selective celiac angiogram, late arterial phase: A densely vascular mass is opacified in the region of the head of the pancreas (**arrowheads**).

B, selective splenic angiogram, arterial phase: Note numerous abnormal vessels in the distal pancreas that have their origin from the splenic artery (**arrowheads**).

(*Continued.*)

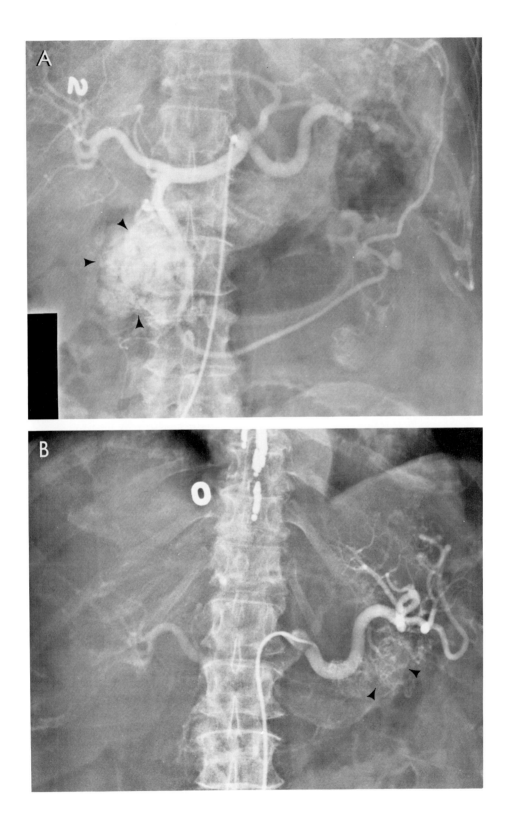

Figure 49 · Metastasis

Figure 49 (cont.).—Metastasis to the pancreas and mesentery from renal cell carcinoma.

C, selective splenic angiogram, venous phase: Tumor blush is seen in the distal pancreas. The splenic vein is partially obstructed in this region. Numerous collateral veins are opacified. The portal vein is well opacified.

D, selective superior mesenteric angiogram, late arterial phase: A nodular lesion with a dense tumor blush is supplied by the ileocolic branch of the superior mesenteric artery in the right lower quadrant (**arrowheads**). A similar smaller lesion is opacified in the left upper quadrant (**arrow**) in the mesentery of the left colon. The tumor blush in the head of the pancreas is present on this injection also.

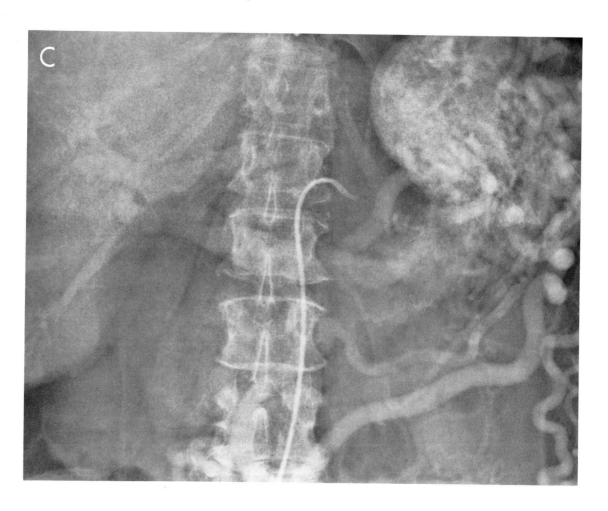

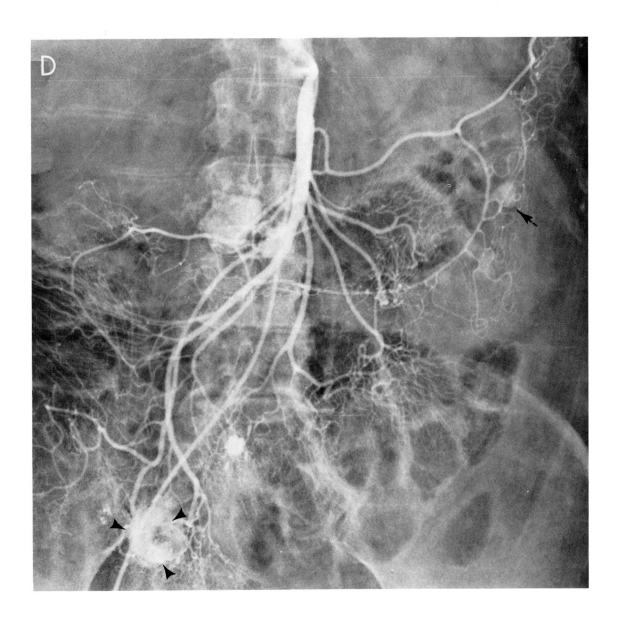

Figure 49 · Metastasis / 117

PART 2

The Liver

Characteristics of Liver Tumors

Primary Malignant Neoplasms

Primary malignant neoplasms of the liver are of two types: tumors of hepatic cell origin are termed hepatomas, and neoplasms of bile duct origin are termed cholangiocarcinomas.

Hepatoma

The occurrence of primary carcinoma of the liver is rare in the United States, constituting less than 2% of digestive tract malignancies. It has, however, a striking geographic and racial variability. This is best exemplified by its occurrence in the Bantu of South Africa, in whom between 40 and 50% of primary malignancies are hepatomas. Nutritional factors have been cited as the cause of this greatly increased incidence among the Bantu.

In the United States, hepatoma affects men more than women in a ratio of 3:1. The tumors occur in 75% of cirrhotic livers. An increased incidence of hepatitis in recent years has been accompanied by a corresponding increase in the incidence of hepatoma. Other etiologic factors of uncertain importance are hemochromatosis and parasitic infestations such as schistosomiasis and clonorchiasis. The greatest frequency of incidence of hepatoma is from the fifth to the seventh decades. A second peak of incidence exists for children under 2 years of age. Four percent of all hepatic malignancies occur before 10 years of age. The majority of those occurring in children are poorly differentiated cellular-type tumors called hepatoblastomas.

There are three gross pathologic types, the most common being multinodular. The capacity of hepatoma to invade veins suggests that the multinodular type represents intrahepatic metastatic spread from a single focus. The second most common form of hepatoma is the massive single focus that most frequently affects the right lobe of the liver. Between 20 and 40% of hepatomas occur in this form. The third type is a diffuse involvement of the entire liver. Histologically, the tumor is composed of cords of liver cells separated by sinusoids. Bile capillaries may be formed; if so, bile stasis will be seen in these capillaries. Large anaplastic giant cells may be included.

Right upper quadrant pain and weight loss are the most common presenting symptoms. Jaundice and bleeding varices are less frequent complaints. On physical examination hepatomegaly is the most frequent finding. Other findings are a definite hepatic mass, splenomegaly and the stigmata of cirrhosis, for example, palmar erythema, testicular atrophy and telangiectasia.

RADIOLOGIC FEATURES.—Hepatomegaly and a right upper quadrant mass are the most frequently seen changes on routine radiologic examination. Many secondary changes caused by hepatic enlargement are also found. These include an elevated right hemidiaphragm and right basilar atelectasis and displacement of the stomach, proximal duodenum, hepatic flexure and right kidney.

Angiography is the most definitive radiologic diagnostic procedure for hepatic neoplasms. Some angiographic changes are nonspecific but occur with all hepatic space-occupying lesions. These include vascular stretching with displacement and increased size of the hepatic artery as well as its main branches as a result of increased functional demand. The characteristic angiographic changes of hepatoma are profuse proliferation of tumor vessels with a dense tumor stain and arteriovenous shunting. The tumor vessels are characterized by their purposeless arrangement, marginal irregularity, saccular dilatation and absence of tapering. Tumor stain occurs in the arterial hepatogram phase of the injection. Within it are numerous small collections of contrast material in amorphous spaces. This phenomenon is termed "puddling." Shunting of some blood from the hepatic artery to the portal vein, less often the hepatic vein, occurs in a high percentage of hepatomas. This is evident from the opacification of a venous structure during the arterial or capillary phase of the hepatic artery injection. Occasionally thrombosis may be demonstrated within a portal vein.

Some hepatomas are of anaplastic cellular type, particularly those occurring in infancy. Many of these will be less vascular than the mature hepatoma and will not have the characteristic angiographic findings. They may appear angiographically similar to cholangiocarcinoma, which is described later. Large hepatomas may outgrow their vascular supply and develop necrotic centers. They may appear as avascular lesions on the angiogram with nonspecific findings of vascular stretching and displacement. Most hepatomas are associated with cirrhosis, and the branches of the hepatic artery not involved by tumor may be abnormal, having the "corkscrew" appearance often found with cirrhosis. Pharmacoangiography, using noradrenalin as the vasoconstricting agent, may be useful to confirm tumor vasculature and staining by the resistance of the tumor vasculature to its action. Vasoconstriction of the normal vessels occurs, and flow to the tumor is increased. This procedure is particularly useful for defining the size of the tumor for operative removal.

CHOLANGIOCARCINOMA

The second primary hepatic neoplasm, cholangiocarcinoma, has the bile duct epithelial cell as its place of origin. This intrahepatic bile duct neoplasm is described here; carcinoma of the extrahepatic biliary ducts is described in Part 3.

The cholangiocarcinoma is an extremely rare tumor, its frequency relative to hepatoma being in a ratio of 1:4. Cholangiocarcinoma occurs twice as frequently in women as in men. Peak incidence is in the fifth, sixth and seventh decades. On the average, the patient with cholangiocarcinoma is slightly older than one with hepatoma. A high percentage of patients with cholangiocarcinoma will have a previous history of cholelithiasis. Thirty-five percent occur in patients having pre-existing cirrhosis.

While the neoplasm is an adenocarcinoma of scirrhous type, the cells resemble those of bile duct epithelium. An abundant fibrous tissue stroma is present.

A diagnosis on clinical grounds is rarely made, as the clinical findings are nonspecific. The most common presenting complaints are right upper quadrant pain and jaundice. Liver enlargement or a mass in the region of the liver may be palpated.

RADIOLOGIC FEATURES.—Findings on the plain radiograph are nonspecific. Liver enlargement or a right upper quadrant mass, elevation of the right diaphragm and a right basilar atelectasis may be evident. The stomach and proximal duodenum, the hepatic flexure of the colon or the right kidney may be displaced if the mass is of sufficient size. Intravenous *cholangiography* demonstrates little of the intrahepatic radicles in the absence of lower biliary tract obstruction. Consequently, positive findings on intravenous cholangiography are a rarity with cholangiocarcinoma. If a T-tube is in place in the common bile duct from a previous operation, a T-tube cholangiogram can be obtained. This examination may demonstrate irregularity and displacement of the intrahepatic biliary tree in the region of the tumor. The changes on the hepatic *angiogram* that occur with cholangiocarcinoma are markedly dissimilar to those of hepatoma. The tumor is poorly vascularized. Pathologic vessels are few and are characteristically thin and beaded. Arterial invasion, encasement and obstruction are the characteristic changes. Vessel proliferation, puddling, tumor stain and arteriovenous shunting are not usually seen. Collateral circulation by arterial anastomosis around the tumor may develop after the arterial encasement or obstruction.

Metastatic Tumors

The most common cause of hepatic mass is metastatic tumor. The gastrointestinal tract is the most common site of origin of the primary neoplasm giving rise to liver metastases, with the colon being the foremost site. Neoplasms that arise in the gastrointestinal tract spread by way of the portal system to the liver. Other intra-abdominal neoplasms, not arising in the gastrointestinal tract, may metastasize to the liver, such as neoplasms of the kidney and uterus. Extra-abdominal primary neoplasms also metastasize to the liver, particularly those of the breast and lung.

RADIOLOGIC FEATURES.—Evidence of metastatic disease to the liver on routine radiographs is minimal. Liver enlargement may be present. *Angiographically,* hepatic metastases can be classified as vascular or avascular, depending on their appearance during the arterial phase and arterial hepatogram phase of the injection. In general, metastases reflect the vascularity of the primary tumor. All metastases appear to be avascular in the portal hepatogram phase as all derive their blood from the hepatic arteries and receive no supply from the portal system. Displacement of the intrahepatic arteries is a common feature.

The vascular category of hepatic metastases is necessarily broad, as it encompasses many gradations of vascularity from moderately vascular to richly vascular (see Table 1).

Vascular metastases rarely show macroscopic tumor vessels but usually show tumor stain in the arterial hepatogram phase. This is the result of contrast material within numerous microscopic tumor vessels. Vascular

TABLE 1—VASCULAR CATEGORIES OF HEPATIC METASTASES

RICHLY VASCULAR

Islet cell pancreatic tumors	Transitional cell carcinoma
Carcinoid tumors	Papillary carcinoma of the pancreas
Leiomyosarcoma	Thyroid carcinoma
Hypernephroma	Choriocarcinoma

MODERATELY VASCULAR

Adenocarcinoma of the breast
Adrenal carcinoma
Seminoma
Occasionally carcinoma of the colon

AVASCULAR

Adenocarcinoma of the pancreas	Wilms' tumor
Adenocarcinoma of the gallbladder	Malignant melanoma
Adenocarcinoma of the bile ducts	Carcinoma of the colon
Carcinoma of the lung	Carcinoma of the esophagus

metastases as small as 1 cm can be demonstrated because of this dense opacification in contrast to the normal surrounding parenchyma. The resolution for avascular metastases is of a lesser degree. Avascular metastases may be difficult to differentiate from other causes of multiple hepatic lucencies, such as cysts and abscesses.

Occasionally, calcification occurs in bulky hepatic metastases. The primary tumors giving rise to these metastases are most commonly mucinous-type adenocarcinomas of the colon and breast. Calcified metastases in the liver are manifested radiographically as fine, discrete densities in poorly defined clusters.

Radionuclide studies of the liver are useful to confirm the presence of, or exclude, hepatic metastases when a primary malignancy has been diagnosed. Serial liver scans or scintillation studies are valuable in following the progress of therapy.

In cases of primary or metastatic liver neoplasms that are undergoing prolonged infusion of chemotherapeutic agents directly into the hepatic arterial circulation by way of an indwelling catheter, injection of contrast agent via the catheter with serial radiographs affords a simple and effective means of following the response of the lesion to treatment. Angiographic assessment of response to treatment is based on reduction in the size of the liver or of the lesion, or both.

Benign Neoplasms

Hemangioma

Hemangiomas consist of two types: the cavernous hemangioma found mostly in adults, and the hemangioendothelioma found most frequently in infants.

Cavernous hemangiomas are found nine times more frequently in women than in men. The incidence is related to multiparity, the tumor being most frequent in older women with four or more pregnancies.

Cavernous hemangiomas may be solitary or multiple. They have no predilection for the left or right lobe of the liver and are usually situated peripherally. The tumor may be extensive and involve an entire lobe or the entire liver. Microscopically, it consists of dilated, epithelium-lined vascular spaces that are supported by fibrous septa. Small vessels communicate with the dilated spaces at the periphery of the tumor. Venous drainage is by way of the hepatic vein, and communication with the portal circulation does not usually exist.

The majority of cavernous hemangiomas of the liver are asymptomatic, are first detected at autopsy or are found incidentally during abdominal arteriography. When symptoms are present, they are mainly the result of pressure on adjoining organs by the enlarged liver. These include pain, nausea, dysphasia, vomiting and jaundice. Intraperitoneal bleeding, melena and hematemesis may occur secondary to rupture. Physical findings are few; a large, firm liver is frequently palpated. A bruit may be heard.

RADIOLOGIC FEATURES.—The routine radiograph of the abdomen, the upper gastrointestinal series, the intravenous pyelogram and the barium enema may show nonspecific signs of liver enlargement or a right upper quadrant mass. An additional rare finding on the plain radiograph is a fine spiculated calcification in a radiating pattern in the liver. This type of calcification may also be rarely found in hepatic metastases from primary carcinoma of the colon.

The *angiographic findings* in hemangioma of the liver are characteristic and allow differentiation of hepatoma and vascular metastases. The intrahepatic arterial branches are normal in appearance to a caliber of 1 mm. Below this size, the arteries are tortuous and fill multiple vascular spaces or lakes which may be round or oval and may be arranged in a circular pattern around the feeding vessel. These vascular spaces fill early, suggesting preferential flow. Contrast material is retained in the fusiform space for up to 20 seconds and then gradually fades. The major intrahepatic arteries will usually be displaced. Arteriovenous shunting rarely occurs in hemangioma, in contrast to hepatoma.

Hemangioendotheliomas affect mostly infants, who may have multiple lesions of this type affecting other organs simultaneously. Calcification, although rare, is more frequent in this tumor than in cavernous hemangioma. The *angiographic pattern* of hemangioendotheliomas differs from that of the cavernous hemangioma in that the feeding arteries are large, and extensive arteriovenous shunting may be seen. Many infants with this condition have congestive heart failure caused by the extensive shunting. Another characteristic feature of the angiographic appearance is lobulation of the abnormal vascular pattern with an avascular center to each lobule.

BENIGN ADENOMA

Benign hepatic adenoma is a rare tumor which may have predominantly hepatic cell or bile duct constitution. A mixed cellular type also occurs. The bile duct adenoma is commonly cystic. The tumor may be single or multifocal.

Clinical features may be right upper quadrant abdominal pain, enlarged liver or palpable liver mass, or the patient may be asymptomatic.

RADIOLOGIC FEATURES.—No characteristic picture is seen on the plain radiograph of the abdomen. One account of an adenoma with angiographic study reported a multifocal lesion causing intrahepatic arterial displacement. There was variable vascularity of the tumor with minimal abnormal tumor vascularity. Pooling of contrast material in central sinusoids in the tumor was demonstrated.

HAMARTOMA

A hamartoma is a rare benign hepatic tumor affecting children and young adults. As with hamartomas affecting other organs, it is composed of disproportionate amounts of the tissues or cells normally occurring in the organ. A similarity exists between this tumor and the mixed type of hepatic adenoma. There is controversy as to whether a distinction can be made between these tumors.

RADIOLOGIC FEATURES.—Angiographically, the tumor appears as a space-occupying lesion with displacement and draping of the intrahepatic arteries. The vascularity is extremely variable. The most frequently described pattern is an avascular mass with few abnormal vessels being observed within the tumor.

SIMPLE CYSTS

Many classifications of simple cysts of the liver have been suggested. A useful classification is difficult to compile because these cysts may be single or multiple, congenital or acquired, or symptomatic or asymptomatic, and it is often impossible to determine an etiologic factor on clinical, radiologic or pathologic examination. Many of these simple cysts are ascribed to aberrations in the development of intrahepatic bile ducts. Blind-end bile ducts may be found in the walls of the cysts. Other simple cysts, termed retention cysts, are secondary to obstructive or static mechanisms in the liver, for example, previous trauma or inflammation.

Simple cysts are rare, are more common in women than in men and occur most frequently in the fourth and fifth decades. The majority, however, are asymptomatic, found incidentally at autopsy.

RADIOLOGIC FEATURES.—A soft tissue mass inseparable from the liver is occasionally seen on the conventional radiograph of the abdomen. Displacement of surrounding organs and structures is similar to that caused by other intrahepatic tumors when the size of the tumor is substantial. With a large cyst, bile duct displacement may be apparent on the intravenous cholangiogram. On angiographic examination, the intrahepatic arteries are displaced by and draped around a totally avascular mass which may be defined clearly with a smooth border distinct from the normal parenchyma during the capillary and portal venous phases. Compression of the contiguous hepatic parenchyma causes increased density in this region. Radionuclide studies show a nonspecific area of decreased uptake conforming to the location of the tumor. The greatest application of this examination is in following the progress of a lesion.

Congenital Polycystic Disease

Fifteen percent of patients having polycystic disease of the kidneys have polycystic disease of the liver. The pancreas may also be involved. Although the condition is congenital, symptoms usually develop in adult life. There is great variability in size and number of the cysts.

Radiologic features.—The liver may be enlarged. Small cysts may not be demonstrable angiographically. Larger cysts appear as avascular mass lesions causing intrahepatic arterial displacement. The cysts will be defined clearly in the capillary and portal hepatogram phases.

Hydatid Cysts

The hydatid cyst is a parasitic infection of the liver caused by the dog tapeworm, *Echinococcus granulosus*. The sheep is the usual intermediate host for the parasite and may be an accidental host. The ova are excreted from the bowel of the dog and as a result of contamination of food or water are ingested by man. They are absorbed from the intestinal tract into the portal vein and are carried to the liver; failing to lodge there, they pass to the lungs by way of the systemic circulation. The embryo develops into a cyst after implantation in an organ.

Hydatid cysts affect the liver in 70% of instances, the majority involving the right lobe. The disease is of worldwide distribution, with particular prominence in the Mediterranean area.

The cyst has three layers. The innermost layer is called the germinal layer, and this gives rise to scolices and daughter cysts. A middle layer, called the endocyst, is the outer layer of the cyst proper and is of a thick hyaline material. The ectocyst is formed by the compressed and scarred tissue of the host organ.

Radiologic features.—In addition to the nonspecific evidence of an intrahepatic mass on the conventional radiograph, calcification may be seen in the right upper quadrant of the abdomen. The calcification is in the wall of the cyst and may be rimlike, circular, semilunar or polycyclic. Similar rimlike calcification in the right upper quadrant may be caused by biliary calculi, renal cysts or adrenal tumors. Major causes of intrahepatic calcification are:

1. Intrahepatic biliary calculi. The calcium may appear as a solid, round density or as rings of radiopaque material.

2. Granulomatous conditions of the liver. The calcification is irregular and amorphous.

3. Metastases from mucin-producing carcinomas, especially of the colon and breast. The calcification may be bulky and dense.

4. Primary hepatic neoplasms. Calcification has been described as occurring in primary hepatic carcinoma, particularly the mixed type. Calcification may also occur in cholangiocarcinoma, cavernous hemangioma and hemangioendothelioma.

The characteristic *angiographic appearance* of hydatid cyst is an avascular intrahepatic lesion with displacement of intrahepatic arteries. A density develops around the avascular mass in the early capillary phase and lasts into the venous hepatogram phase. This dense circumferential blush may be the result of hypervascularity of the compressed liver tissue that comprises the ectocyst. The rim is well defined in its inner margin; its outer margin gradually fades and merges with the normal hepatogram.

HEPATIC ABSCESS

Infective organisms can be introduced into the liver by various routes. The most common route is the portal vein. Emboli from suppurative appendicitis or other infective processes of the bowel are carried by the mesenteric veins to the portal vein and then to the liver. The resulting abscesses are usually multiple and most commonly involve the right lobe. The most common causative organisms are *Escherichia coli* and streptococci. Septic emboli from pyogenic bacterial endocarditis may reach the liver through branches of the hepatic artery and may form scattered small abscesses throughout the liver. Infection may ascend into the liver parenchyma if cholangitis caused by *E. coli* is present. A perforated intestinal viscus may give rise to a subdiaphragmatic abscess, and infection may spread from there to the liver. Infection may be introduced to the liver by penetrating trauma or operation.

In amebic dysentery, amebas from the ulcers in the bowel wall are carried by the portal system to the liver and cause multiple abscesses. Apart from histologic identification of the causative organism, these abscesses are indistinguishable from those caused by pyogenic organisms. In general, the former are larger and of longer duration.

Actinomycotic infection caused by the fungus *Actinomyces bovis* may also be carried to the liver by the portal system. This organism also causes multiple liver abscesses.

RADIOLOGIC FEATURES.—Observations on the conventional radiograph vary from no visible abnormality to gross liver enlargement. A localized mass protruding from the superior border of the liver is occasionally outlined by the lucency of the lung above. An air-fluid level in the liver may be seen on a radiograph of the abdomen with the patient upright. Secondary inflammatory change may involve the pleura or the lung. The angiographic appearance of an abscess in its early acute stage is simply that of an avascular intrahepatic mass causing displacement of intrahepatic arteries. The avascular lesion may be indistinct or poorly defined. At a later stage of development of the abscess, abnormal vascularization develops because of granulation tissue. The abnormal vessels surround the radiolucent center. In the arterial hepatogram phase a rim of density is formed around the radiolucent center, giving a halo effect. The rim is wide, irregular and poorly defined. The abnormal vessels may simulate tumor vascularity but, in conjunction with the radiolucent center, should suggest an abscess, although necrotic primary neoplasms or metastases could have a similar appearance.

SUGGESTED READING

Anderson, J. E., and Perlmutter, G. S.: Diagnosis of hepatoma using a multiple radionuclide approach, Radiology 102:387–389, February, 1972.

Baum, S.: Hepatic Arteriography, Chapter 61, in Abrams, H. L. (ed.): *Angiography* (2nd ed.; Boston: Little, Brown and Company, 1971), Vol. 2, pp. 983–1002.

Bieler, E. U.; Meyer, B. J., and Jansen, C. R.: Liver scanning as a method for detecting primary liver cancer: Report of 100 cases, Am. J. Roentgenol. 115:709–716, August, 1972.

Boijsen, E., and Abrams, H. L.: Roentgenologic diagnosis of primary carcinoma of the liver, Acta Radiol. 3:257–277, May, 1965.

Caplan, L. H., and Simon, M.: Nonparasitic cysts of the liver, Am. J. Roentgenol. 96:421–428, February, 1966.

Colapinto, R. F.: Arteriography in the diagnosis of liver tumors, Canad. M. A. J. 99: 1175–1185, December 21, 1968.

Farrell, R.; Steinman, A., and Green, W. H.: Arteriovenous shunting in a regenerating liver simulating hepatoma: Report of a case, Radiology 102:279–280, February, 1972.

Jewel, K. L.: Primary carcinoma of the liver: Clinical and radiologic manifestations, Am. J. Roentgenol. 113:84–91, September, 1971.

Kaude, J., and Rian, R.: Cholangiocarcinoma, Radiology 100:573–580, September, 1971.

Kido, C.; Sasaki, T., and Kaneko, M.: Angiography of primary liver cancer, Am. J. Roentgenol. 113:70–81, September, 1971.

Kreel, L.; Jones, E. A., and Tavill, A. S.: A comparative study of arteriography and scintillation scanning in space-occupying lesions of the liver, Brit. J. Radiol. 41: 401–411, June, 1968.

McLoughlin, M. J.: Angiography in cavernous hemangioma of the liver, Am. J. Roentgenol. 113:50–55, September, 1971.

McNulty, J. G.: Angiographic manifestations of hydatid disease of the liver: A report of two cases, Am. J. Roentgenol. 102:380–383, February, 1968.

Moss, A. A., et al.: Angiographic appearance of benign and malignant hepatic tumors in infants and children, Am. J. Roentgenol. 113:61–69, September, 1971.

Nebesar, R. A.; Pollard, J. J., and Stone, D. L.: Angiographic diagnosis of malignant disease of the liver, Radiology 86:284–292, February, 1966.

Palubinskas, A. J.; Baldwin, J., and McCormack, K. R.: Liver-cell adenoma: Angiographic findings and report of a case, Radiology 89:444–447, September, 1967.

Pantoja, E.: Angiography in liver hemangioma, Am. J. Roentgenol. 104:874–879, December, 1968.

Pollard, J. J.; Fleischli, D. J., and Nebesar, R. A.: Angiography of hepatic neoplasms, Radiol. Clin. North America 8:31–41, April, 1970.

Pollard, J. J.; Nebesar, R. A., and Mattoso, L. F.: Angiographic diagnosis of benign diseases of the liver, Radiology 86:276–283, February, 1966.

Rizk, G. K.; Tayyarah, K. A., and Ghandur-Mnaymneh, L.: The angiographic changes in hydatid cysts of the liver and spleen, Radiology 99:303–309, May, 1971.

Rossi, P., and Ruzicka, F. F., Jr.: Differentiation of intrahepatic and extrahepatic masses by arteriography, Radiology 93:771–780, October, 1969.

Sackett, J. F., et al.: Scintillation scanning of liver cell adenoma, Am. J. Roentgenol. 113:56–60, September, 1971.

Shonfeld, E. M.; Guarino, A. V., and Bessolo, R. J.: Calcified hepatic metastases from carcinoma of the breast: Case report and review of the literature, Radiology 106:303–304, February, 1973.

Stulberg, J. H., and Bierman, H. R.: Selective hepatic arteriography: Normal anatomy, anatomic variations, and pathological conditions, Radiology 85:46–55, July, 1965.

Viamonte, M., Jr.; Warren, W. D., and Fomon, J. J.: Liver panangiography in the assessment of portal hypertension in liver cirrhosis, Radiol. Clin. North America 8:147–167, April, 1970.

Watson, R. C., and Baltaxe, H. A.: The angiographic appearance of primary and secondary tumors of the liver, Radiology 101:539–548, December, 1971.

William, B.: *Pathology for the Physician* (7th ed.; Philadelphia: Lea and Febiger, 1965).

Williams, R. C., and Wise, R. E.: Infusion hepatic angiography: Assessment of hepatic malignancy via the infusion catheter, Radiol. Clin. North America 8:43–51, April, 1970.

Yü, C: Primary carcinoma of the liver (hepatoma): Its diagnosis by selective celiac arteriography, Am. J. Roentgenol. 99:142–149, January, 1967.

Figure 50.—Hepatoma.

A 46-year-old man complained of pain in the right upper quadrant of the abdomen. Physical examination revealed a palpable mass.

A, selective celiac angiogram, arterial phase: Note the enlarged common hepatic artery (**arrows**) with displacement and draping of its branches caused by a large vascular mass in the inferior portion of the right lobe of the liver. Profuse tumor vascularity and arteriovenous shunting are demonstrated in the tumor (**arrowheads**). The stomach and colon are displaced inferiorly and to the left by the large tumor. The gastroduodenal artery is visible to the left of the midline.

B, selective celiac angiogram, late arterial phase: Note the venous opacification.

(*Continued.*)

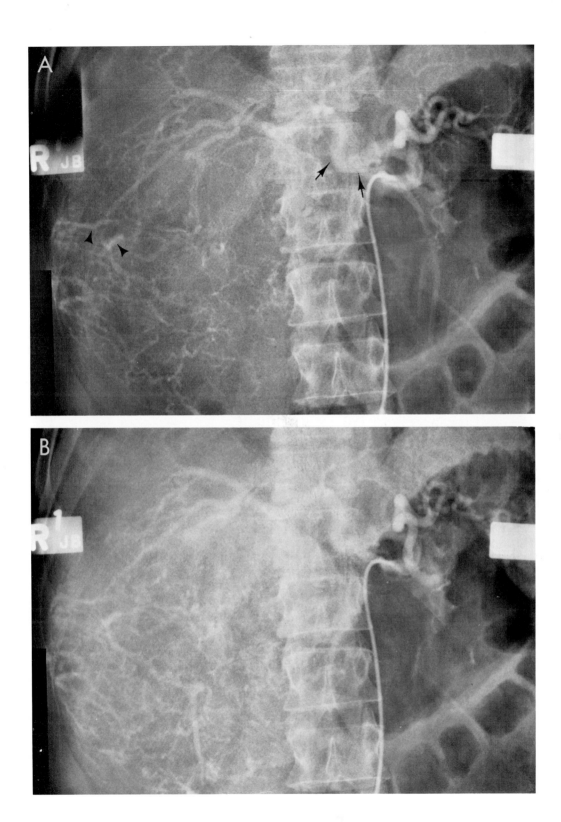

Figure 50 · Hepatoma

Figure 50 (cont.).—Hepatoma.

C, selective celiac angiogram, portal venous phase: A tumor blush outlines the large neoplasm (**arrowheads**).

D, liver scan, anteroposterior projection: Note deficient uptake in the lower portion of the right lobe in the region of the tumor.

Comment: This is the uninodular type of hepatic cell carcinoma. This case demonstrates the angiographic findings most characteristic of hepatoma, for example, large feeding arteries, profuse tumor vascularity, contrast medium puddling and arteriovenous shunting.

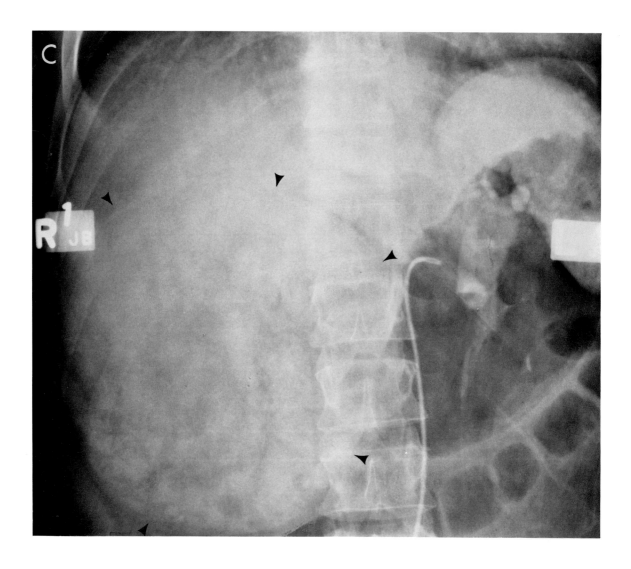

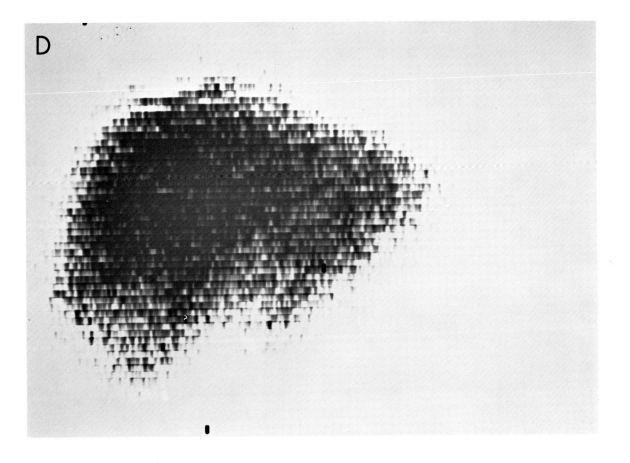

Figure 50 · Hepatoma

Figure 51.—Hepatoma.

This man was hospitalized for emergency investigation and treatment of acute upper gastrointestinal bleeding.

A, selective celiac angiogram, arterial phase: Note displacement and draping of the intrahepatic arteries in both the right and the left lobe of the liver. Tumor vessels are evident throughout the liver. This tumor is moderately vascular. Some contrast medium puddling is seen within the tumor.

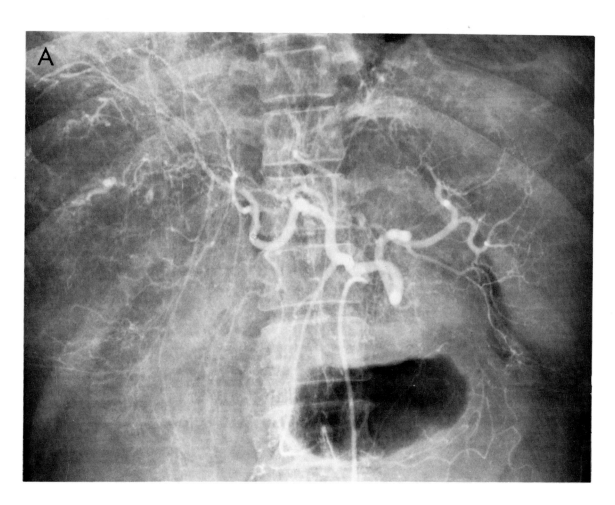

B, selective celiac angiogram, early hepatogram phase: The diffuse involvement of the entire liver with abnormal tumor vessels is evident. Some arteriovenous shunting is visible. The relatively avascular appearance of the tumor in the lateral margin of the right lobe may be caused by tumor necrosis.

(Continued.)

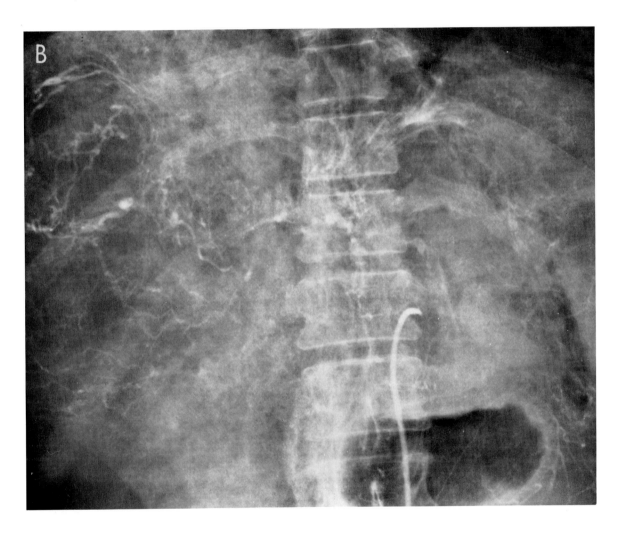

Figure 51 (cont.).—Hepatoma.

C, selective superior mesenteric angiogram, arterial phase: A superior mesenteric artery injection shows displacement of the inferior pancreaticoduodenal arteries and the midcolic artery medially by the enlarged right lobe of the liver (**arrowheads**).

D, selective superior mesenteric angiogram, venous phase: Narrowing and partial obstruction of the portal vein at the hilus of the liver may be seen (**arrows**). There is retrograde flow of contrast material into the splenic vein and inferior mesenteric vein. Collateral venous channels in the region

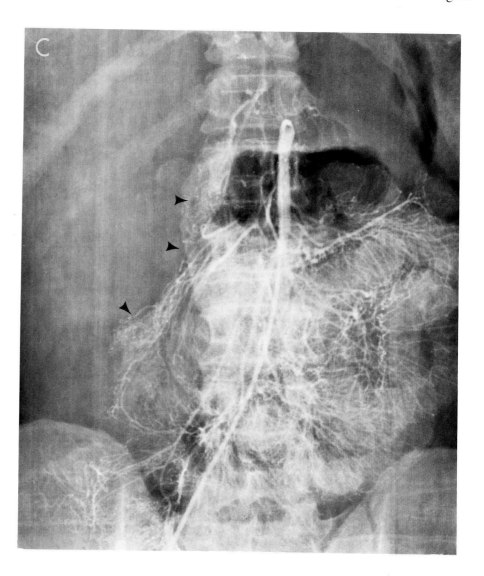

of the fundus of the stomach are varicose (**arrowheads**). A filling defect within the portal vein in apposition with its inferior margin may represent thrombus.

Comment: This case exemplifies the diffuse type of hepatoma. There is widespread involvement of both lobes of the liver. The relative avascularity of the tumor in the lateral aspect of the right lobe of the liver probably represents tumor necrosis.

Figure 51, courtesy of Dr. S. Baum, Massachusetts General Hospital, Boston.

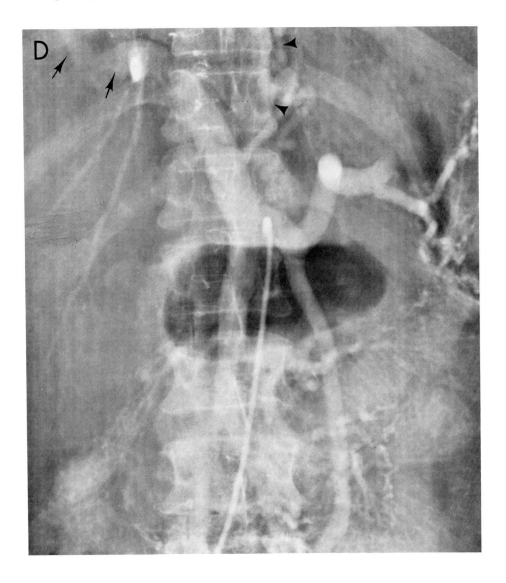

Figure 52.—Hepatoma.

A 20-year-old man had a gradual onset of painless jaundice.

A, upper gastrointestinal radiograph, anteroposterior projection: A soft tissue mass which is roughly circular in outline is present in the upper portion of the abdomen overlying the lower border of the right upper lobe of the liver (**arrowheads**). An amorphous calcification is present within the lesion. There is medial displacement of the first and second portions of the duodenum.

B, selective common hepatic angiogram, arterial phase: The inferior branches of the right hepatic artery which are supplying the richly vascular hepatic mass are minimally enlarged. The mass contains many irregular tumor vessels, and puddling of contrast material is evident. The position of the tumor, lying superficially and protruding from the inferior margin of the right lobe of the liver, accounts for the absence of obvious displacement of the intrahepatic arteries.

(*Continued.*)

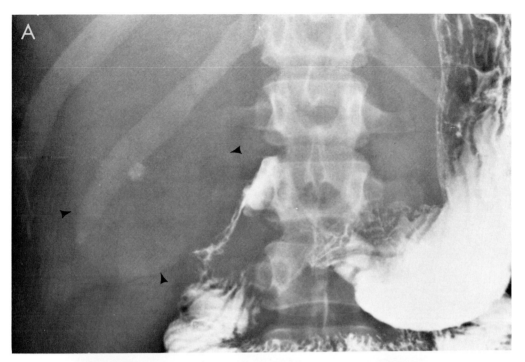
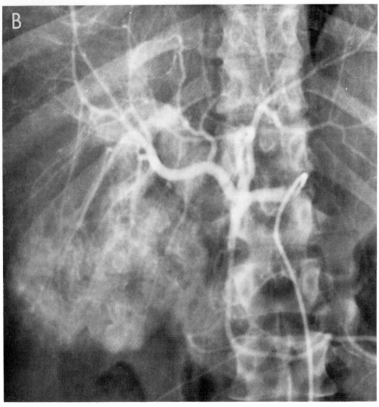

Figure 52 · Hepatoma

Figure 52 (cont.).—Hepatoma.

C, selective common hepatic angiogram, early hepatogram phase: A pronounced tumor blush outlines the mass. Some veins are outlined by contrast material as a result of arteriovenous shunting within the neoplasm (**arrowheads**).

D, selective common hepatic angiogram, arterial phase: This study was done 16 months after operation for removal of the hepatoma of the right lobe. The left lobe of the liver shows compensatory hypertrophy. Localized narrowing and some lateral displacement of the gastroduodenal artery are the result of the operative procedure. There is no evidence of recurrence of the hepatoma.

Comment: This is the uninodular type of hepatoma. The tumor is well circumscribed and localized to a portion of the right lobe. This type of tumor is amenable to surgery. Calcification within a hepatoma is rare, although a number of cases have been described. Primary hepatic carcinoma of mixed cell type is more prone to calcification than is hepatoma.

Figure 52, courtesy of Dr. S. Baum, Massachusetts General Hospital, Boston.

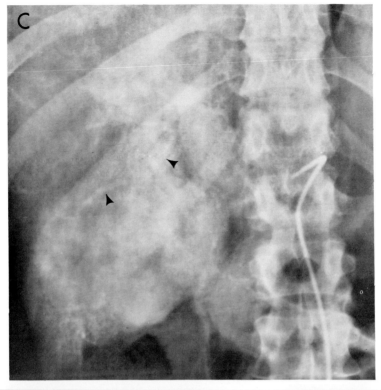

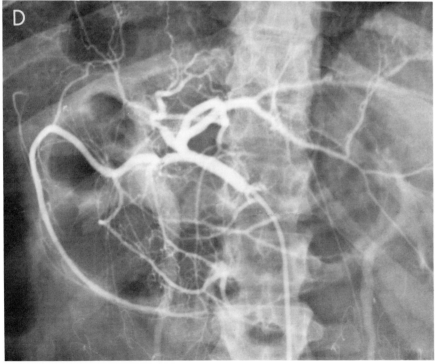

Figure 52 · Hepatoma

Figure 53.—Hepatoma.

A 69-year-old woman complained of anorexia, weight loss, nausea and vomiting and abdominal pain. A duodenal ulcer was diagnosed by upper gastrointestinal examination. During operation for partial gastrectomy, multiple tumor nodules were palpated in the liver. Biopsy results revealed hepatoma. An indwelling infusion catheter for continuous prolonged chemotherapy was placed in the common hepatic artery at a later date.

Infusion hepatic angiogram, capillary phase: Note multiple vascular lesions throughout the greatly enlarged liver. Those involving the left lobe are well defined; a larger lesion in the right lobe is poorly defined.

Comment: This is the multinodular type of hepatoma. Controversy exists as to whether this form represents local metastasis to the liver from what was initially a single tumor. As hepatoma has a propensity for invasion of veins, it is quite conceivable that local spread of this type could occur.

The infusion hepatic angiogram is obtained by injection into the infusion catheter at a rate of 2 cc/second using a mechanical injector. Approximately 75 cc of contrast material is injected. The manual film-changing technique is used, and five to seven radiographs are obtained. This is an excellent method of evaluating response to chemotherapy.

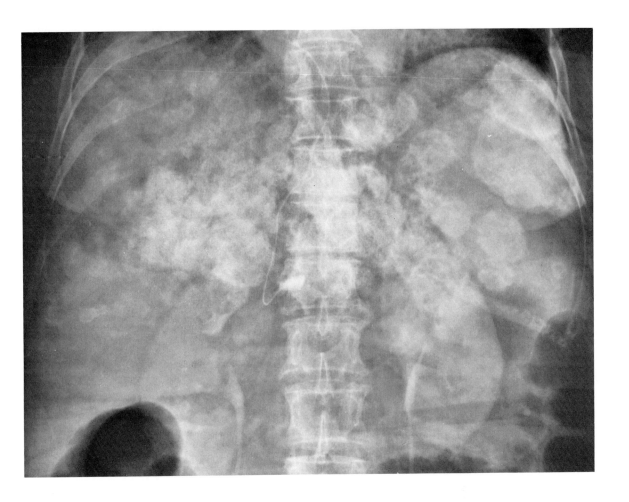

Figure 53 · Hepatoma / 145

Figure 54.—Hepatoma with spindle cell metaplasia.

A 24-year-old woman was examined soon after a normal pregnancy and delivery because of chest pain, abdominal distention and hepatomegaly.

A, selective celiac angiogram, arterial phase: There is gross displacement of the right and left hepatic arteries by a large avascular mass involving both right and left lobes of the liver. The intrahepatic arteries are draped over the mass (**arrowheads**). The liver is grossly enlarged, in particular the left lobe. Some abnormal vasculature is evident at the border of the neoplasm, particularly on the left. Minimal vascular puddling is noted in this

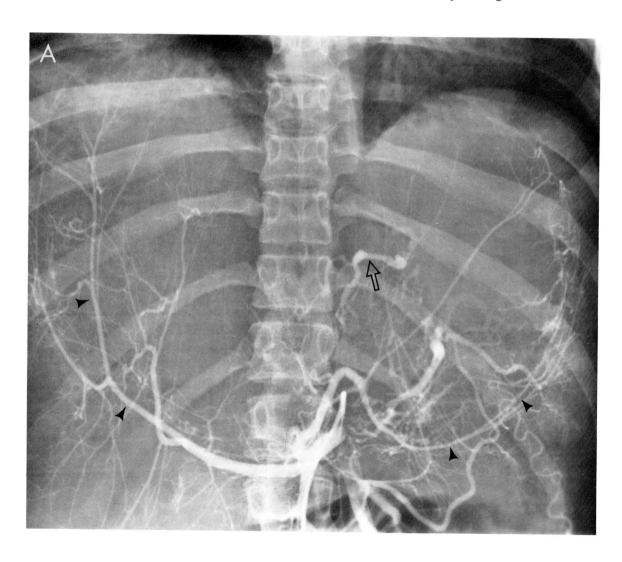

same region. An unusual vascular structure is beginning to fill (**arrow**); this overlies the proximal end of the eleventh rib on the left and has the appearance of a vein.

B, selective celiac angiogram, early hepatogram phase: Note the avascular nature of the large intrahepatic tumor. The periphery of the tumor is ill defined and mottled. There is early filling of the veins within the left lobe. The largest vein, a large collateral vein, courses superiorly and passes beyond the confines of the liver. Its appearance at this phase of the angiogram is the result of arteriovenous shunting.

(*Continued.*)

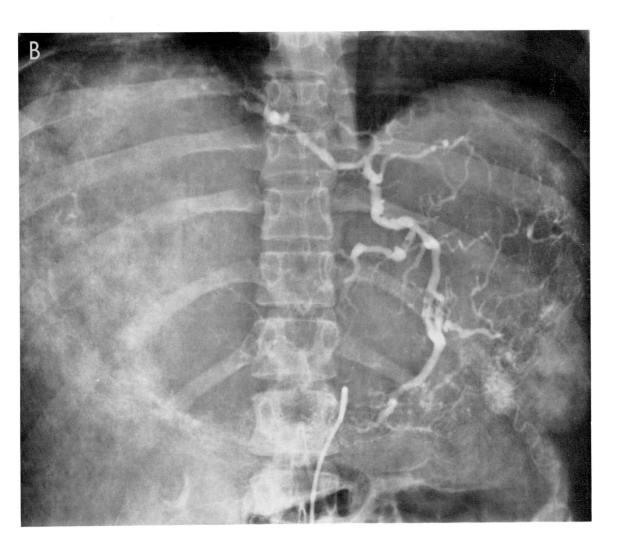

Figure 54 (cont.).—Hepatoma with spindle cell metaplasia.

C, selective celiac angiogram, portal venous phase: The collateral vein that filled by way of the arteriovenous anastomosis from the left hepatic artery (**B**) is diminishing in opacification. The portal hepatogram is deficient in the region of the tumor. The left branch of the portal vein is not visible.

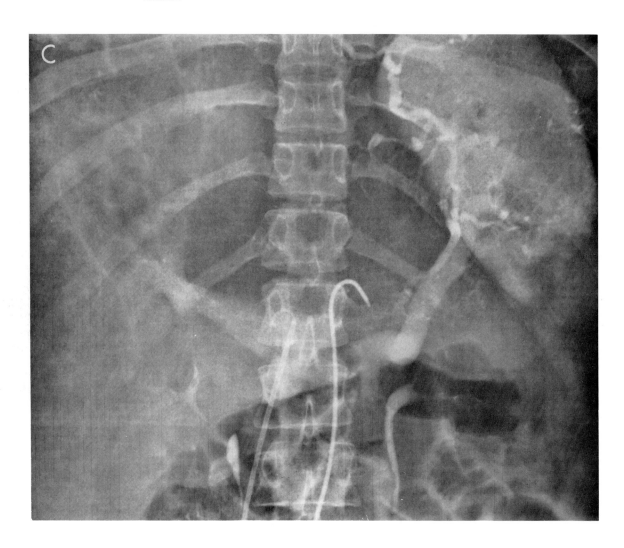

D, liver scan, anteroposterior projection: There is a large area of diminished uptake involving the right and left lobes corresponding to the tumor mass demonstrated angiographically.

Comment: The pathologic diagnosis of hepatoma with spindle cell metaplasia indicates a sarcomatous appearance to this tumor. This type of primary hepatic tumor is extremely rare. The avascularity of the neoplasm is unusual for hepatoma and probably reflects the sarcomatous change in this tumor. Some of the characteristics of hepatoma are retained, for example, arteriovenous shunting.

Figure 54, courtesy of Dr. S. Baum, Massachusetts General Hospital, Boston.

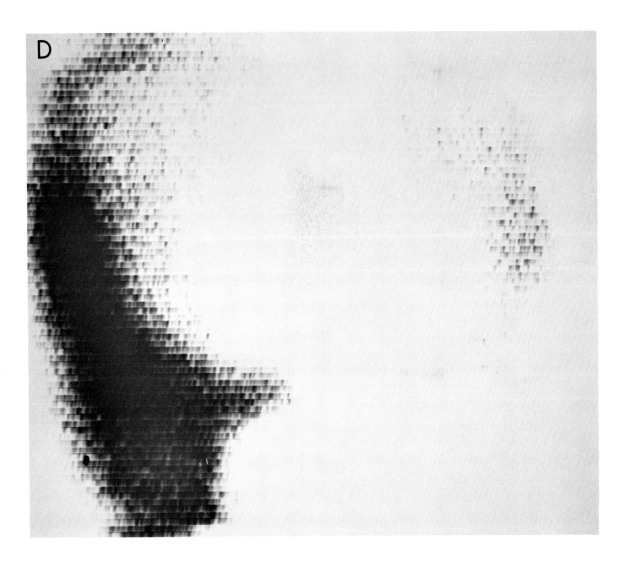

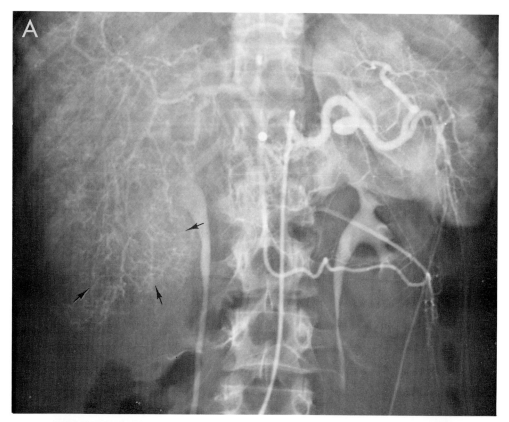
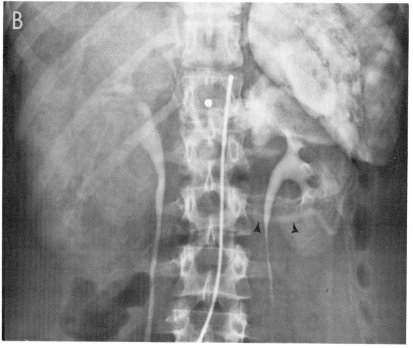

150 / The Liver

Figure 55.—Hepatoma.

A 13-year-old child complained of right upper abdominal pain of sudden onset. The liver was enlarged to 8 cm below the right costal margin and was tender on palpation.

A, selective celiac angiogram, arterial phase: The branches of the hepatic artery to the lower border of the right lobe are enlarged. Abnormal tumor vessels are delineated in this portion of the liver (**arrows**). The tumor is ill defined. No arteriovenous shunting is demonstrated in this case.

B, selective celiac angiogram, portal venous phase: The splenic vein is completely obstructed near the midline. A collateral vein is seen running transversely at the level of the second lumbar vertebra (**arrowheads**). The neoplasm has the appearance of an avascular defect in the liver.

C, liver scan, anteroposterior projection: A large area of deficient uptake occupies most of the right lobe of the liver. The left lobe is enlarged, possibly as a result of compensatory hypertrophy.

Comment: Partial or complete obstruction of the portal vein is an occasional complication of hepatoma. The obstruction may simply be caused by compression or may be caused by tumor invasion and thrombus. Demonstration of a hepatic tumor with early delineation of a portal vein which contains thrombus is said to be pathognomonic of hepatoma.

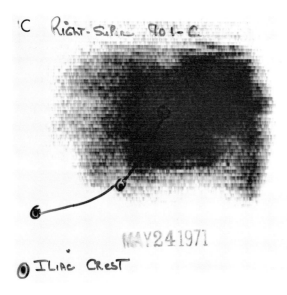

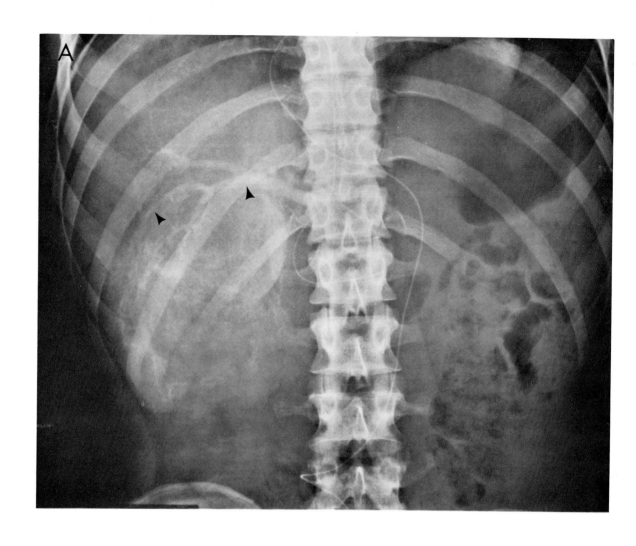

Figure 56.—Hepatoma.

An 18-year-old youth complained of right-sided abdominal pain for 10 days. He had been feeling ill for some time and had intermittent nausea and vomiting. Palpation revealed a tender mass in the right upper quadrant of the abdomen.

A, infusion hepatic angiogram, late arterial phase: This film was obtained after injection of a small volume of contrast material through the infusion catheter. A tumor of mixed vascularity is demonstrated in the lower and medial border of the right lobe of the liver. There are displacement and draping of the branches of the right hepatic artery in this region (**arrowheads**).

B, liver scan, anteroposterior supine projection: A large area of decreased uptake is present in the lower medial border of the right hepatic lobe (**arrows**), corresponding to the site and size of the neoplasm seen on angiography.

Comment: The lower and medial portion of the right lobe of the liver is the commonest site of origin of intrahepatic neoplasms. Some difficulty in interpretation of radionuclide studies of the liver is the result of physiologic defects in this region, most notably caused by the gallbladder and the hilus of the liver.

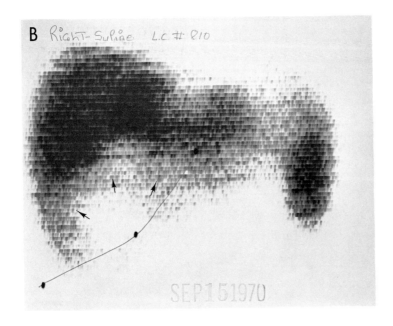

Figure 57.—Hepatoma.

A 73-year-old woman had a 20-lb. weight loss and complained of anorexia, nausea, vomiting and abdominal fullness with discomfort. The liver was palpable 12 cm below the right costal margin.

A, selective celiac angiogram, arterial phase: The intrahepatic arteries are displaced; they are crowded in the superior portion of the right lobe and are separated inferiorly. A minimal amount of abnormal vascularity is seen scattered in the right lobe of the liver. Vascular puddles and arteriovenous shunts are not demonstrated in this case.

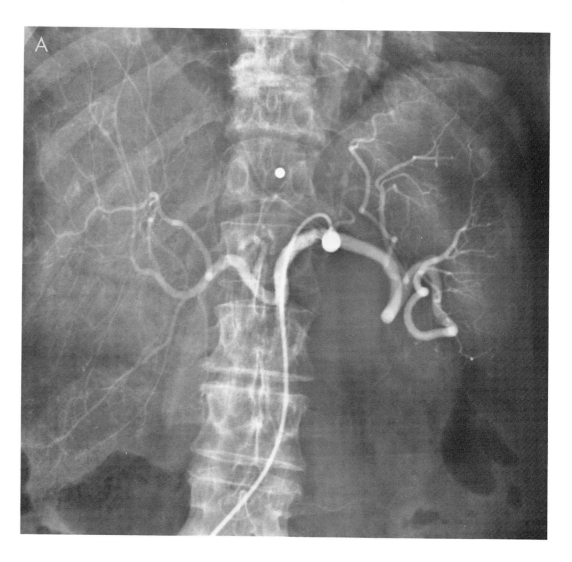

B, selective celiac angiogram, portal venous phase: The portal hepatogram is nonhomogeneous, with the inferior portion of the right lobe of the liver presenting a mottled appearance. A branch of the portal vein to this region of the liver is obstructed (**arrow**). A lucent defect is present in the superolateral section of the right lobe.

Comment: This case does not exhibit many of the angiographic characteristics of hepatoma. This may be the result of necrosis within the tumor, particularly in the superolateral section of the right lobe. Confirmation of this impression could not be obtained from the pathologist's report.

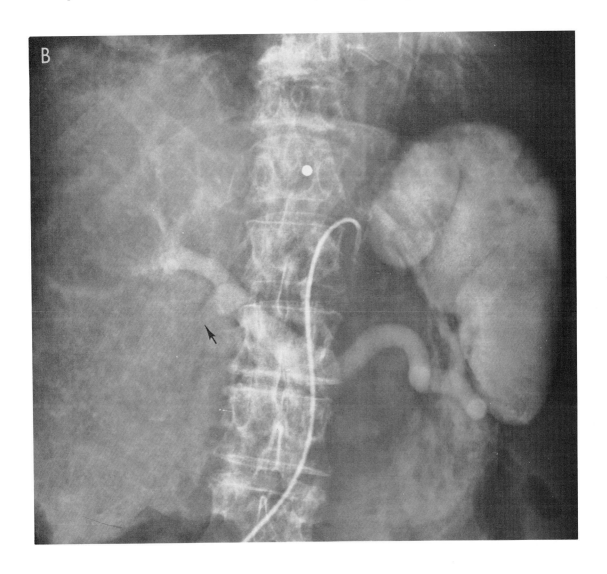

Figure 58.—Cholangiocarcinoma.

This woman complained of increasing right upper quadrant pain over a period of six weeks. Jaundice and chills had developed three days before hospitalization. During operation for cholecystectomy and common duct exploration, a mass was noted in the hilus of the liver.

A, T-tube cholangiogram, spot film, right anterior oblique projection: Marked narrowing and irregularity involve the bifurcation of the common hepatic duct and the proximal portions of the right and left hepatic ducts (**arrowheads**). The right hepatic duct appears to be dilated proximal to the stenosis. A small amount of contrast medium passes beyond the stricture to the left hepatic duct.

B, selective celiac angiogram, arterial phase: Note the localized area of irregularity, tortuosity and narrowing of the arteries within the distribution of the left hepatic artery (**arrowheads**).

C, selective celiac angiogram, portal venous phase: The portal vein is narrowed immediately superior to the position of the T-tube (**arrow**) and corresponds to the location of the stenosis noted on the T-tube cholangiogram (**A**).

Figure 58, courtesy of Dr. S. Baum, Massachusetts General Hospital, Boston.

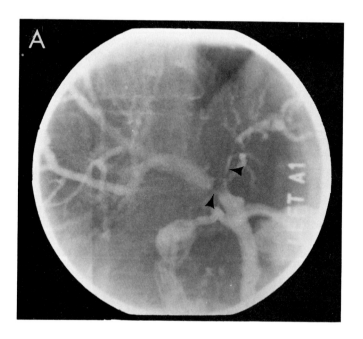

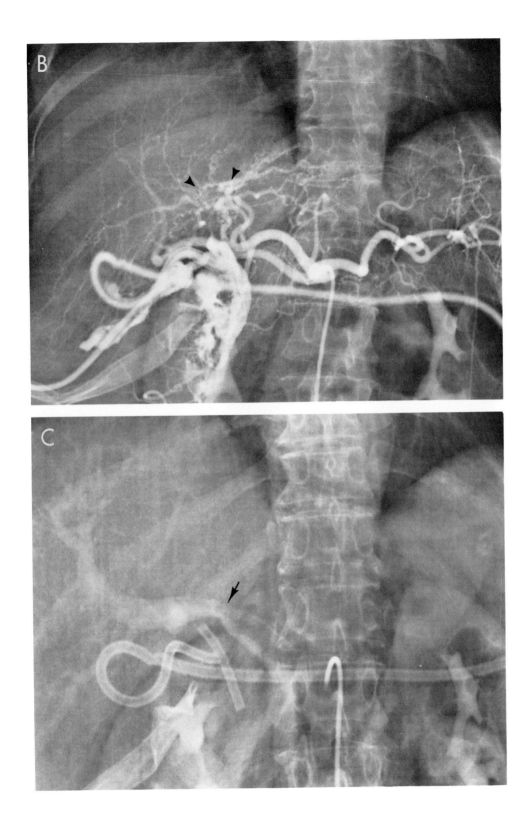

Figure 58 · Cholangiocarcinoma / 157

Figure 59.—Hepatic metastasis from primary carcinoma of the breast.

A, selective celiac angiogram, arterial phase: A large mass which is moderately vascular overlies the midabdomen. The celiac axis and its major branches and the right hepatic and the gastroduodenal arteries are displaced by the mass. The gas shadow of the stomach shows it is likewise displaced to the left.

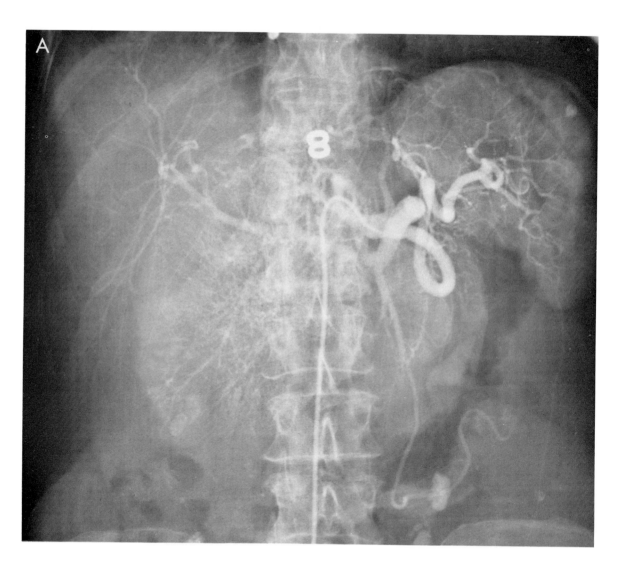

B, selective celiac angiogram, portal venous phase: The large circular mass is again outlined. Impression of the contiguous portion of the right hepatic lobe is evident. The splenic and portal veins are partly compressed by the tumor, but complete obstruction has not occurred.

Comment: This metastasis is unusual in that it is single and has grown largely outside the bounds of the liver.

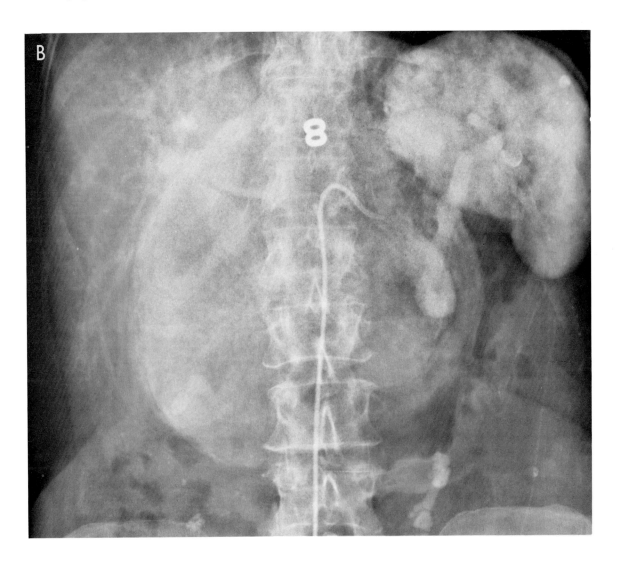

Figure 60.—Hepatic metastasis from adenocarcinoma of the pancreas.

Selective celiac angiogram, venous phase: Large well-defined nodules with moderately intense tumor blush remain opacified long after the arterial phase of the injection. The splenic vein is completely obstructed as a result of the pancreatic carcinoma. The collateral veins are delineated in the left abdomen (**arrowheads**).

Comment: A small number of pancreatic carcinomas, mainly those of the capillary type, produce metastases in the liver which are more vascular than the surrounding parenchyma.

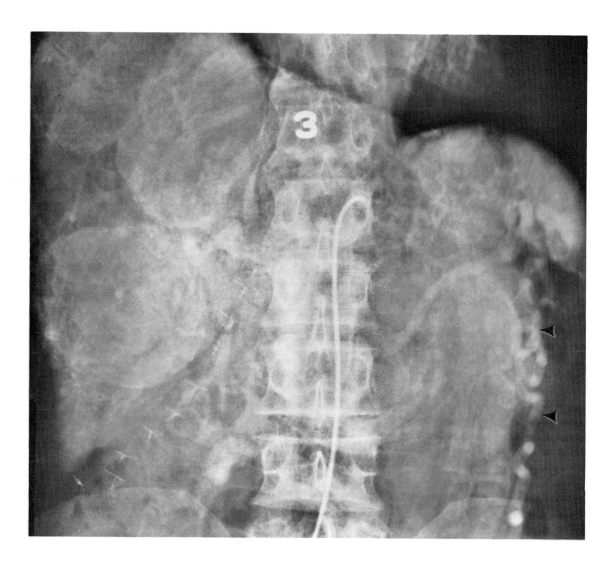

Figure 61.— Hepatic metastasis from carcinoid tumor.

A 20-year-old man complained of flushing and palpitation, abdominal cramps and diarrhea after eating. He had a grade 2 systolic murmur heard maximally in the pulmonary valve area. The liver was tender and enlarged. Results of a liver biopsy confirmed the diagnosis of carcinoid tumor.

Infusion hepatic angiogram, arterial phase: The right hepatic artery fills from this injection by way of the indwelling infusion catheter. Numerous circular, circumscribed hypervascular lesions are seen within the right lobe of the liver.

Comment: The hepatic metastases of carcinoid tumors are most frequently of a highly vascular type. The left hepatic lobe is not visible in this instance, as it was being infused by way of a Kifa catheter.

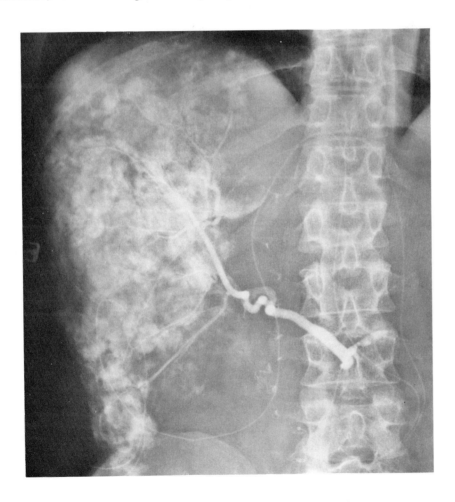

Figure 62.—Hepatic metastases from primary adenocarcinoma of the breast.

A 64-year-old woman had had a left radical mastectomy seven years before this study. She now complained of anorexia and nausea. Her liver was enlarged, extending 12 cm below the right costal margin, and was nodular and tender.

Infusion hepatic angiogram, arterial phase: The left lobe of the liver is massively enlarged. Numerous highly vascular lesions are scattered throughout both lobes of the liver. The intrahepatic arteries are grossly deformed and displaced. There are nonopacified segments of the liver, particularly in the right lobe, possibly representing vascular obstruction or tumor necrosis in this region. The dense opacification of the celiac, common hepatic and proximal portions of the proper hepatic and gastroduodenal arteries represents a partial intramural injection of contrast material.

Comment: Both vascular and avascular intrahepatic neoplasms primary or metastatic to the liver are susceptible to central necrosis. In some hepatic metastases, angiography reveals vascular and avascular foci probably due to necrosis within nodular lesions.

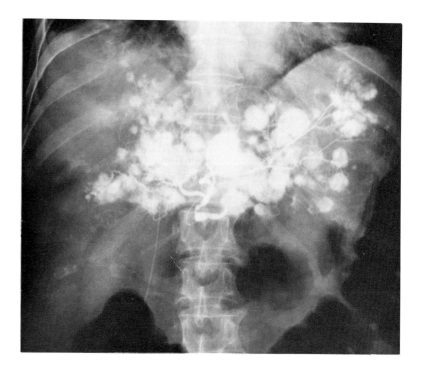

Figure 63.—Hepatic metastasis from primary carcinoma of the colon.

A 39-year-old woman had a carcinoma of the colon diagnosed three previous to this examination. At the time of operation, metastases to the liver and regional lymph nodes were noted. The angiographic pattern of these metastases is mixed, some being hypervascular and some avascular.

Infusion hepatic angiogram, arterial phase: The right and left hepatic arteries and their branches are greatly irregular in caliber and distribution. The early hepatogram is extremely mottled; many areas of the liver appear to have hypervascular-type lesions. Numerous avascular lesions are also scattered in the liver.

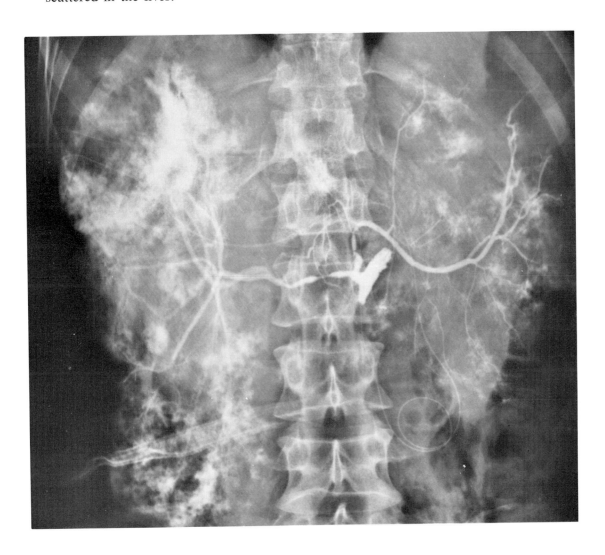

Figure 64.—Hepatic metastasis from primary carcinoma of the colon.

A 43-year-old woman had a carcinoma of the colon resected and was found to have multiple hepatic nodules.

A, selective superior mesenteric angiogram, arterial phase: The right hepatic artery arises from the superior mesenteric artery. This is a variation that is seen in 11% of cases. The distal intrahepatic arteries are minimally displaced, with small avascular intrahepatic lesions situated laterally and superiorly in the right lobe (**arrows**).

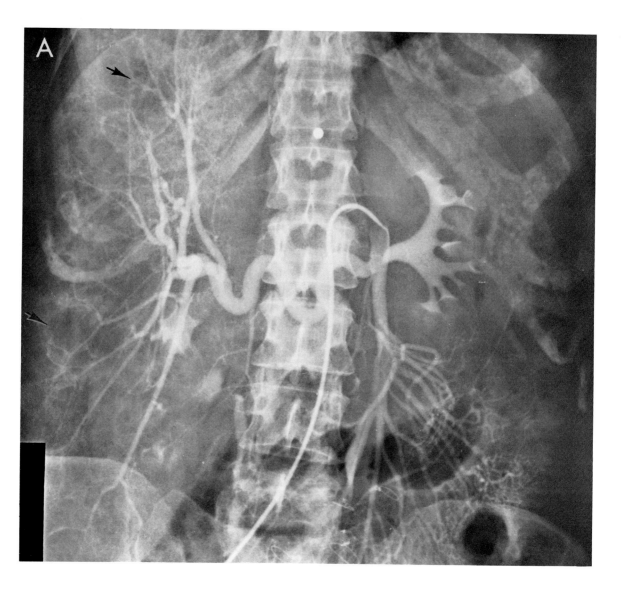

B, infusion hepatic angiogram, late phase: The avascular metastatic lesions situated peripherally in the right lobe of the liver are again demonstrated on this later examination following infusion of a chemotherapeutic agent.

(*Continued.*)

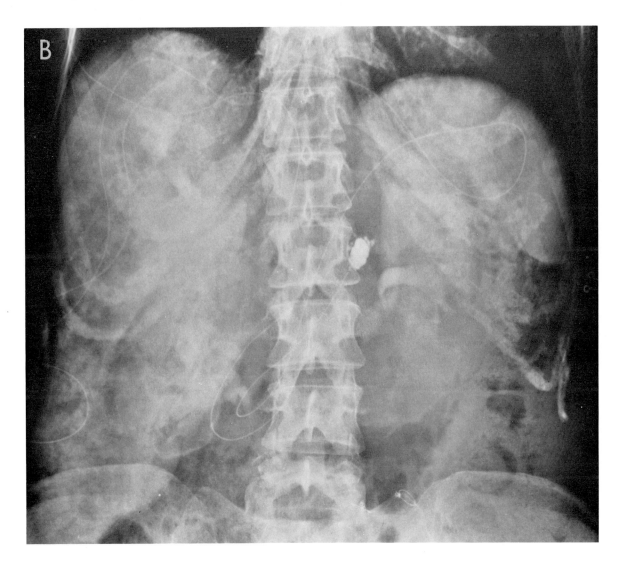

Figure 64 · Metastasis from Colon

Figure 64 (cont.).—Hepatic metastasis from primary carcinoma of the colon.

C, liver scan, anteroposterior supine projection: The liver is greatly enlarged. Numerous areas of decreased uptake are seen scattered throughout both lobes, corresponding to the avascular lesions seen on the angiographic studies.

D, liver scan, right lateral projection: The areas of decreased uptake are again demonstrated.

Comment: The liver scan is used in conjunction with infusion hepatic angiography for continuing evaluation of the response of hepatic metastases to infusion of chemotherapeutic agents. The liver scan is useful in ruling out metastasis in patients with known primary carcinomas. The scan can resolve lesions more than 2 cm in diameter.

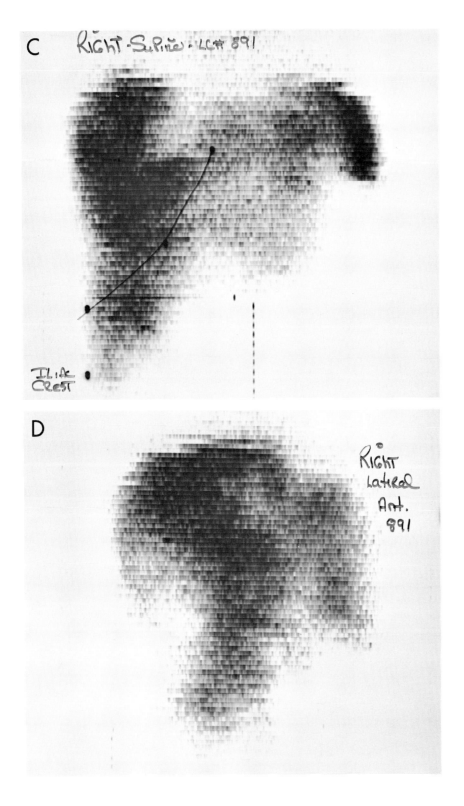

Figure 64 · Metastasis from Colon / 167

Figure 65.—Hepatic metastasis from primary carcinoma of the colon.

A 61-year-old man had carcinoma of the descending colon resected six months previously. Liver metastases were noted at the time of operation.

A, selective celiac angiogram, arterial phase: There are draping and displacement of the smaller intrahepatic arterial branches within the lower portion of the right lobe of the liver (**arrows**). The vessels are displaced around an avascular mass in this portion of the liver.

B, liver scan, anteroposterior projection: The liver is greatly enlarged. Large areas of deficient uptake are evident in the right lobe of the liver.

C, liver scan, right lateral projection: Two regions of decreased perfusion are evident. The first is within the inferior portion of the right lobe; the second is situated posteriorly and superiorly in the right lobe.

Comment: Most hepatic metastases from primary carcinomas of the colon are avascular.

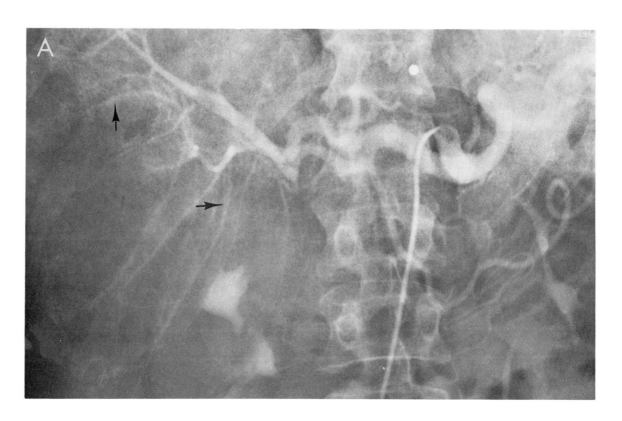

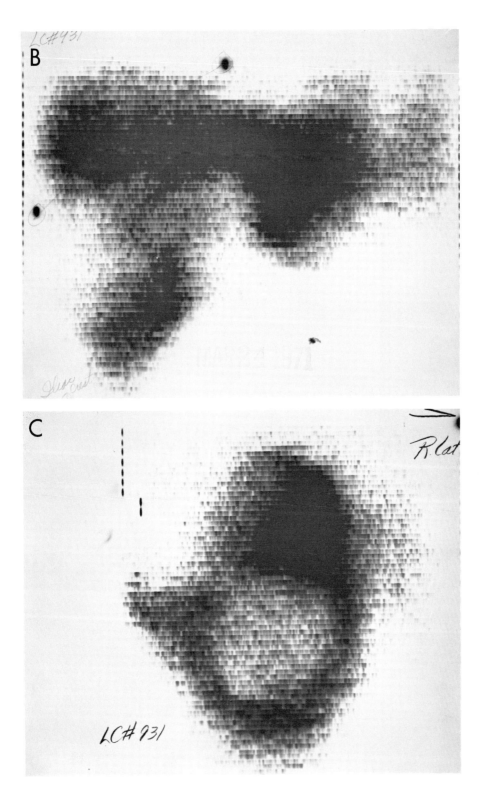

Figure 65 · Metastasis from Colon / 169

Figure 66.—Hepatic metastasis from primary carcinoma of the colon.

A 39-year-old woman had a carcinoma of the colon diagnosed three months previously. Liver metastases were noted at the time.

Portal venous infusion angiogram: An infusion catheter is situated in the portal vein. The left branch of the portal vein is less well perfused than the right. The right portal hepatogram shows three large, well-defined lucencies that represent the metastatic lesions.

Comment: Infusion of the portal vein has been supplanted by hepatic artery infusion. Hepatic metastases receive the major portion of their blood supply from the hepatic artery.

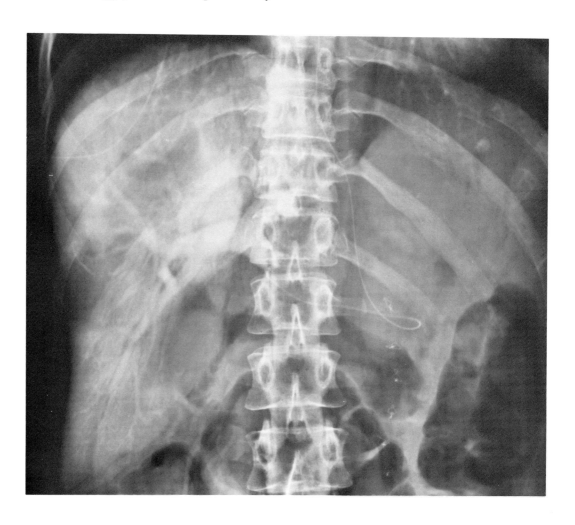

Figure 67.—Hepatic metastasis from primary carcinoma of the colon.

A 64-year-old woman had a left colectomy for carcinoma performed two years before this examination. The liver is now palpable to 8 cm below the right costal margin.

Infusion hepatic angiogram, arterial phase: The proper hepatic artery, a portion of the gastroduodenal artery and the right hepatic artery are delineated. The intrahepatic arterial branches on the right are grossly distorted and displaced. The early hepatogram shows numerous vascular and avascular lesions within the right lobe, which is greatly enlarged.

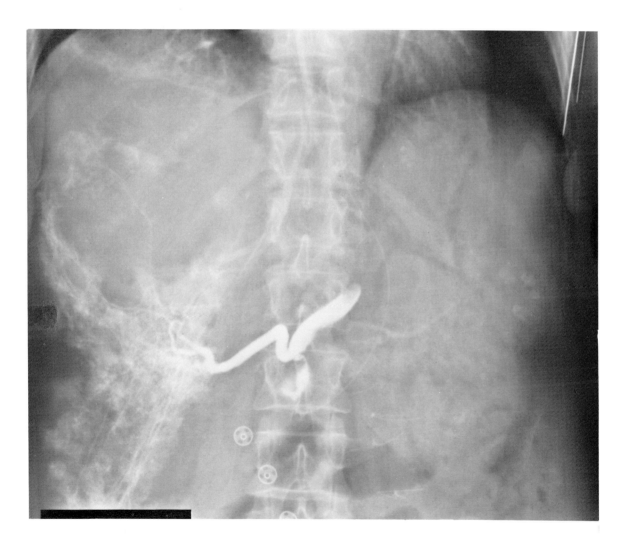

Figure 68.—Hepatic metastasis from primary carcinoma of the colon.

A 66-year-old woman had a carcinoma of the colon resected two years previously; at that time, metastases in the liver were noted.

A, portal venous infusion angiogram: The right branch of the portal vein is obstructed; the left branch of the portal vein is well demonstrated. The left lobe of the liver is greatly enlarged. The early portal hepatogram is mottled.

B, portal venous infusion angiogram: Numerous well-circumscribed lucencies are seen within the portal hepatogram.

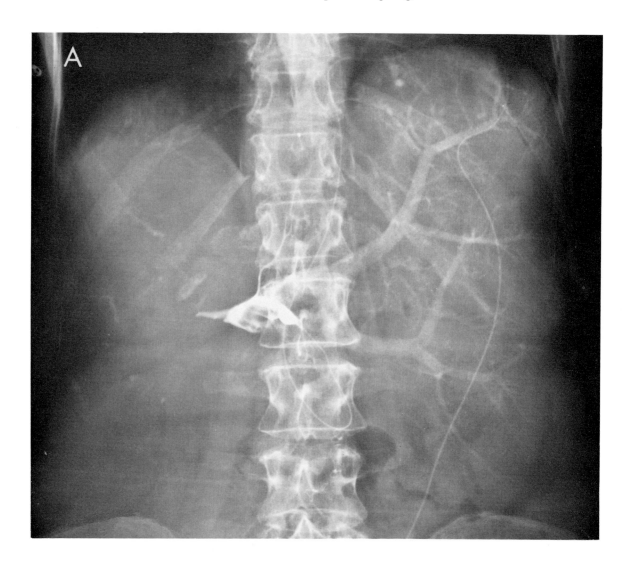

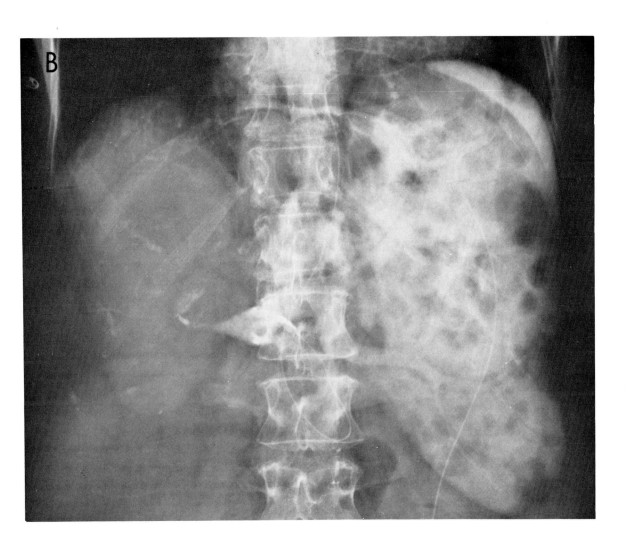

Figure 68 · Metastasis from Colon

Figure 69.—Hepatic metastasis from primary carcinoma of the colon.

A 60-year-old woman initially complained of lower abdominal pain, diarrhea and rectal bleeding. Carcinoma of the rectosigmoid was diagnosed on sigmoidoscopy.

A, selective celiac angiogram, arterial phase: The right lobe of the liver is greatly enlarged. There are draping and attenuation of the branches of the right hepatic artery around avascular lesions within the right lobe (**arrows**).

B, infusion hepatic angiogram, hepatogram phase: Numerous avascular defects in the hepatogram represent metastatic foci.

(*Continued.*)

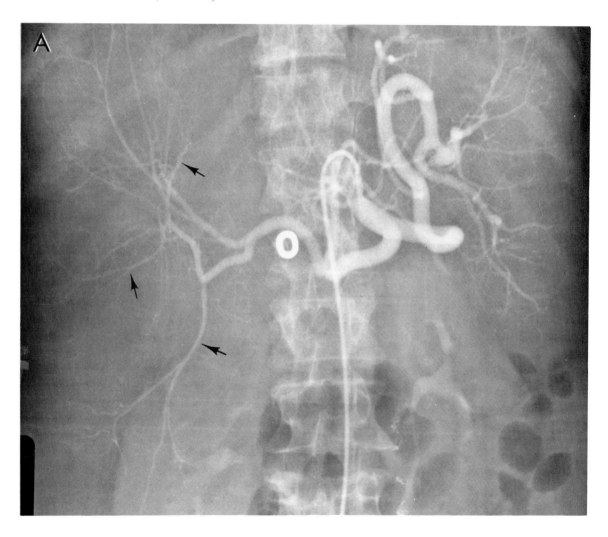

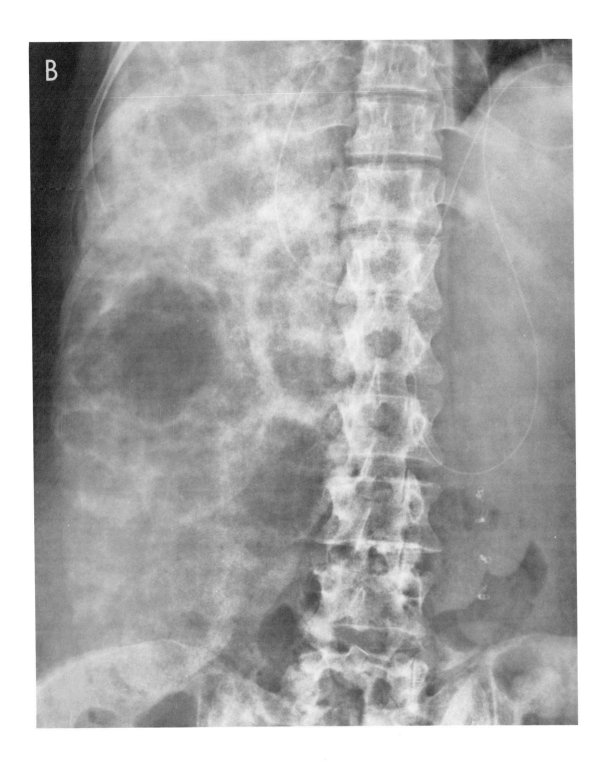

Figure 69 · Metastasis from Colon

Figure 69 (cont.).—Hepatic metastasis from primary carcinoma of the colon.

C, liver scan, anteroposterior supine projection: The liver is massively enlarged. Multiple areas of deficient uptake of the radionuclide represent the metastatic lesions.

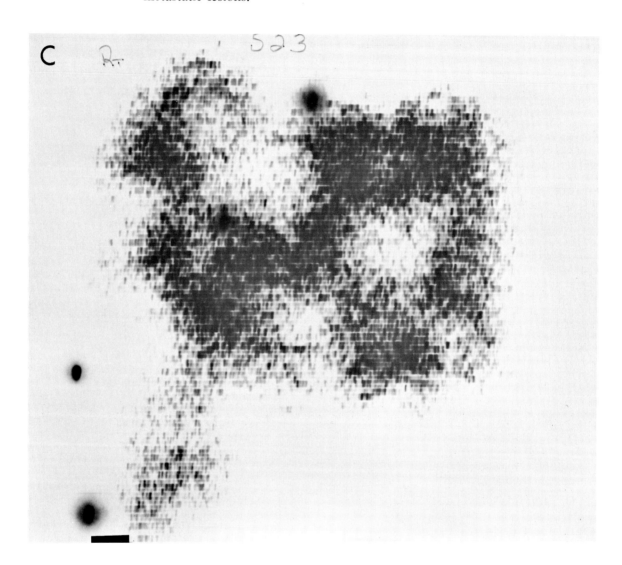

Figure 70.—Hepatic metastasis from leiomyosarcoma of the jejunum.

This elderly woman had a leiomyosarcoma of the jejunum removed 1 year before this examination. Carcinoma of the breast had been diagnosed and a simple mastectomy had been performed 11 months before. The liver became enlarged and a superficial mass was palpable in the lower abdomen. Biopsy of both revealed a leiomyosarcoma.

Infusion hepatic angiogram, hepatogram phase: A large vascular mass is present within the right lobe of the liver (**arrowheads**). The central portion of this mass is more radiolucent than the surrounding region. The appearance is that of a vascular lesion with some central necrosis.

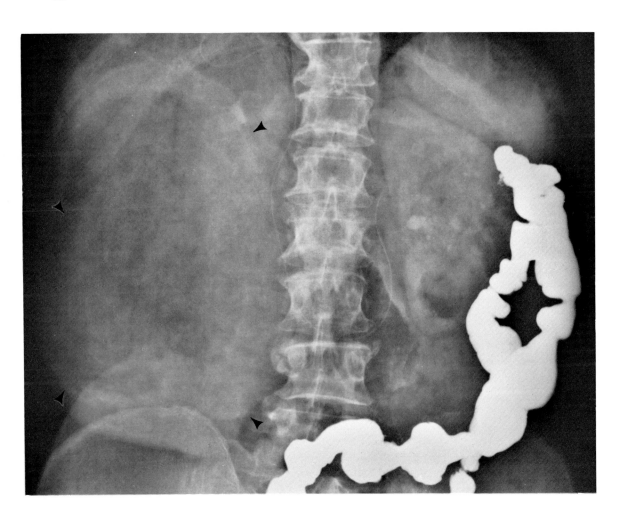

Figure 70 · Metastasis from Jejunum

Figure 71.—Hepatic metastasis from primary fibrosarcoma of the lower limb.

A 52-year-old woman had had a low-grade fibrosarcoma excised from the right thigh 12 years previously.

A, infusion hepatic angiogram, arterial phase: The liver is greatly enlarged. The arterial branches in both lobes (**arrowheads**) are displaced. There are numerous avascular filling defects within the liver.

B, infusion hepatic angiogram, venous phase: The hepatic veins are visible within the right and left lobes. Two large avascular lesions are clearly outlined in the right lobe and at least two in the left lobe.

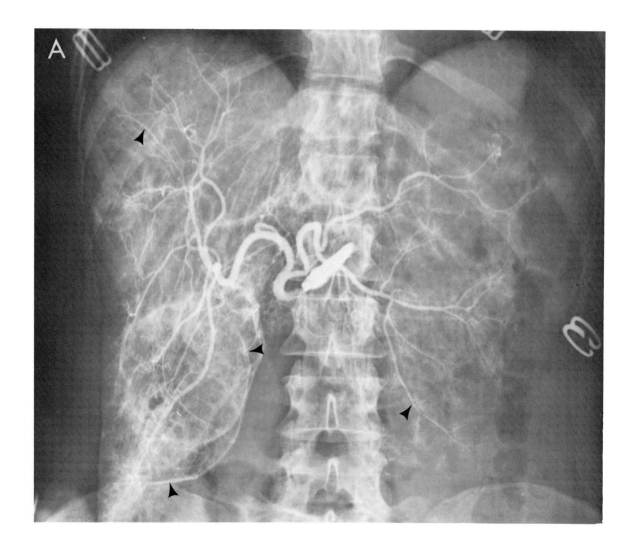

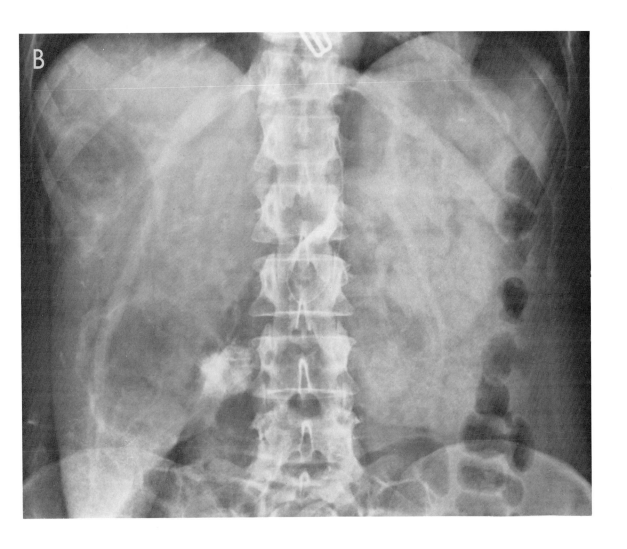

Figure 71 · Metastasis from Fibrosarcoma / 179

Figure 72.—Cavernous hemangioma of the liver.

In a 42-year-old man, multiple filling defects had first shown on a liver scan eight years previously. There was a recent onset of jaundice and hepatic enlargement.

A, selective superior mesenteric angiogram, arterial phase: The right hepatic artery arises from the superior mesenteric artery. This anatomic variation occurs in 11% of cases. The branches of the right hepatic artery are displaced, separated and draped. The proximal branches of the right

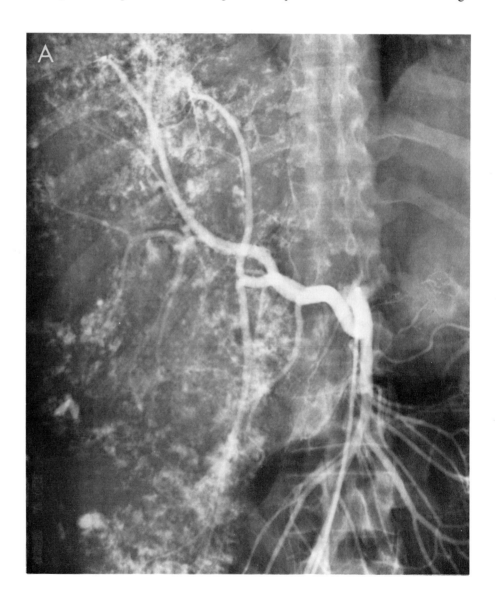

hepatic artery appear to be quite large. The more distal intrahepatic arteries appear to be normal in caliber. Numerous small sinusoidal collections of contrast material are scattered throughout the right lobe of the liver.

B, selective superior mesenteric angiogram, late phase: The sinusoids are still visible in the right lobe of the liver, and these were in evidence 30 seconds after the beginning of the injection. There is no evidence of arteriovenous shunting.

(Continued.)

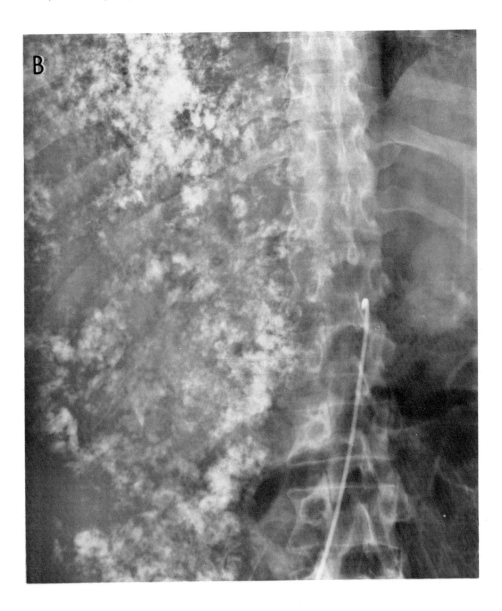

Figure 72 (cont.).—Cavernous hemangioma of the liver.

C, selective celiac angiogram, arterial phase: The celiac axis and its branches are displaced to the left by the mass on the right lobe of the liver. The position of the gastroduodenal artery, to the left of the midline (**arrow**), suggests that the duodenum is also displaced to the left. The left lobe of the liver is enlarged. The branches of the left hepatic artery appear to be normal.

D, selective celiac angiogram, portal venous phase: The portal vein is displaced to the left (**arrows**). The left branch of the portal vein is visible, but the right branch is obstructed. Some minimal involvement of the left

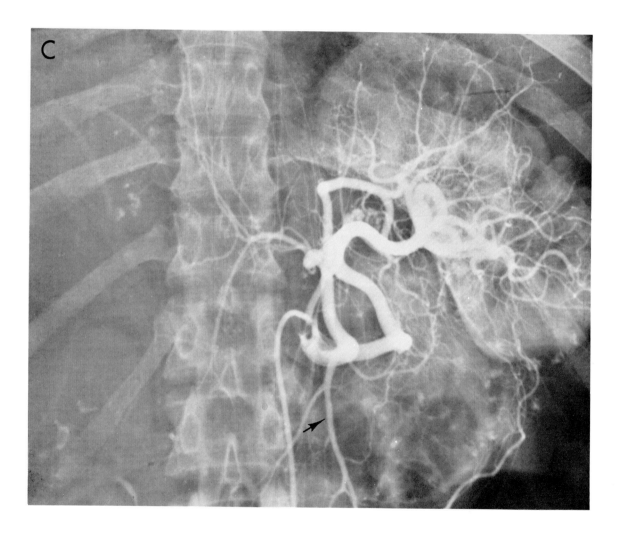

lobe with hemangioma is now visible, as contrast puddling is evident immediately superior to the midsplenic vein.

Comment: Some characteristic angiographic features of cavernous hemangiomas of the liver are demonstrated in this case. The normal caliber of the distal intrahepatic arteries, delayed retention of contrast material in vascular sinusoids, or puddles, and the absence of arteriovenous shunting are typical findings in hemangioma of the liver. The almost complete involvement of the right lobe is unusual. Portal vein obstruction is also uncommon. The enlargement of the right lobe had also caused bile duct obstruction and jaundice.

Figure 72, courtesy of Dr. S. Baum, Massachusetts General Hospital, Boston.

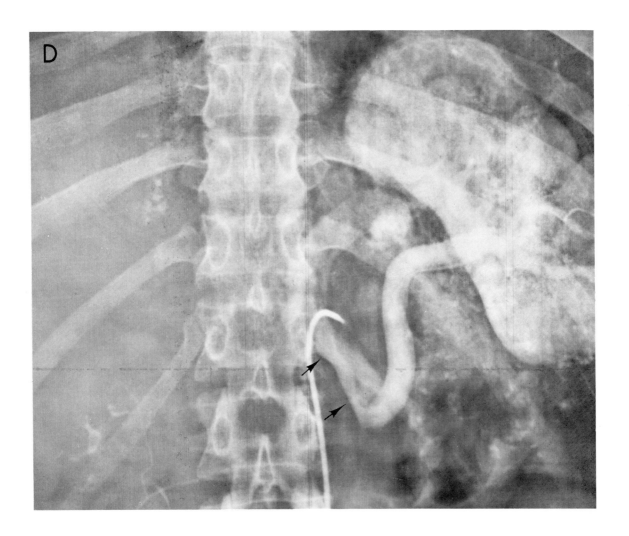

Figure 73.—Cavernous hemangioma of the liver.

A 75-year-old woman complained of weight loss, nausea and anorexia. Physical examination revealed an enlarged liver.

A, selective superior mesenteric angiogram, arterial phase: The right hepatic artery arises from the superior mesenteric artery. Minute irregular vessels are opacified in the tip of the right lobe of the liver. There is filling of vascular sinusoids or lacunae in this region (**arrowheads**). A venous structure is beginning to fill in the region of the lesion (**arrow**). A second, much smaller lesion is faintly visible, lying superiorly and laterally in the right lobe.

(Continued.)

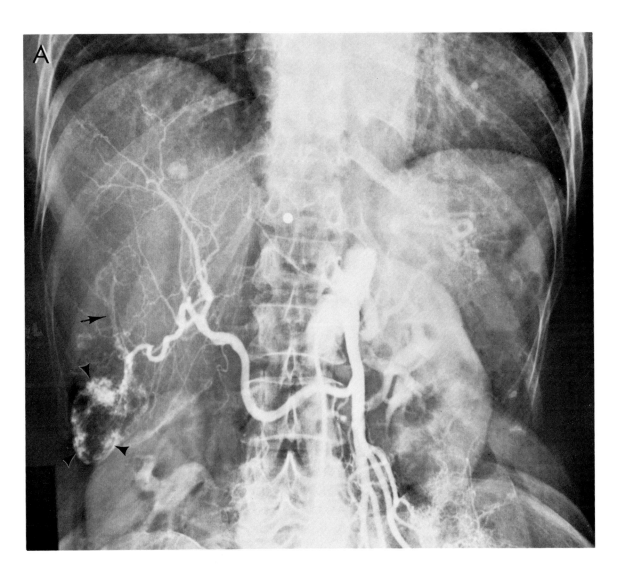

Figure 73 · Cavernous Hemangioma / 185

Figure 73 (cont.).—Cavernous hemangioma of the liver.

B, selective superior mesenteric angiogram, arterial hepatogram phase: The contrast material is retained in numerous small vascular spaces in the lesion in the right lobe of the liver (**arrowheads**). These spaces are not uniformly distributed. The venous structure is now more clearly identified (**open arrow**). The second and similar lesion, much smaller in size, is visible superiorly in the right lobe (**arrow**).

C, selective celiac angiogram, portal venous phase: The venous structure that was seen to fill in the early stage of the arterial hepatogram is now identified as a smaller branch of the right branch of the portal vein (**arrow**). Some filling of the vascular sinusoids within the tumor has occurred.

Comment: Arteriovenous shunting is an infrequent angiographic finding in cavernous hemangioma. This phenomenon is more characteristic of hepatoma. Minimal abnormal tumor vessels are visible in this case, and they are sometimes seen with hemangioma. The retention of contrast material in puddles within the vascular sinusoids is typical.

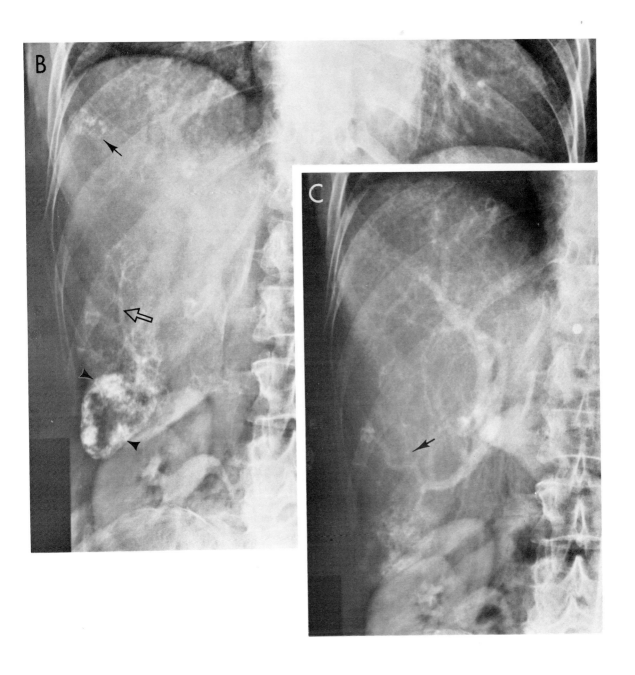

Figure 73 · Cavernous Hemangioma / 187

Figure 74.—Hamartoma of the liver with cystic degeneration.

A 15-year-old girl had a palpable mass protruding from beneath the right costal margin that had been increasing in size over the previous 18 months.

A, selective celiac angiogram, arterial phase: Note the smooth draping and displacement of the branches of the right hepatic artery around a mass in the inferior border of the right lobe of the liver (**arrowheads**). The cystic artery is also displaced inferiorly and medially. Few vessels are delineated within the lesion. Tumor blush is not visible. The appearance is that of an avascular hepatic mass.

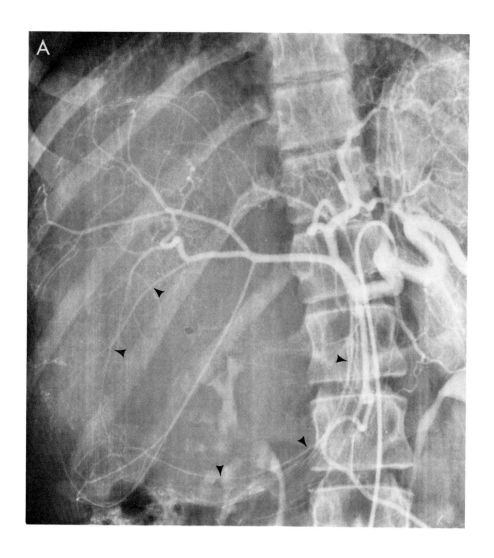

B, selective celiac angiogram, serial magnification study of right hepatic arteries: A few abnormal tumor vessels are identified by this magnification technique (**arrows**). They are tortuous, have a purposeless distribution and exhibit a lack of tapering and normal arborization. Vascular sinusoids, tumor stain and arteriovenous shunting are not visible.

Comment: The identification of tumor vasculature indicates a neoplastic lesion rather than a cyst. The tumor is for the most part avascular, and this coincides with the pathologic finding of cystic degeneration within the tumor. There is wide variation in the described angiographic findings in hamartomas.

Figure 74, courtesy of Dr. S. Baum, Massachusetts General Hospital, Boston.

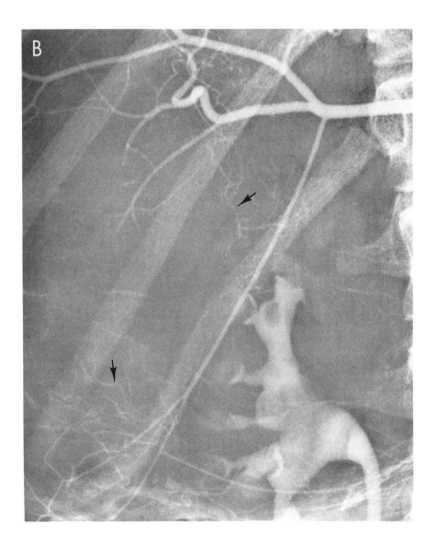

Figure 75.—Liver cell adenoma.

A 22-year-old woman had mild asymptomatic hypertension and had gained 15 lb. The liver was enlarged and blood pressure measured 170/100 mm Hg.

A, aortogram, late arterial phase: There are draping and stretching of the secondary arterial branches around three separate masses within the right lobe of the liver (**arrows**). Many abnormal tumor vessels arise from these arteries and course within the liver.

B, aortogram, late phase using magnification technique: Note an unusual reticular pattern of veins in the central right hepatic mass (**arrows**). These retain the contrast material longer than the surrounding parenchyma.

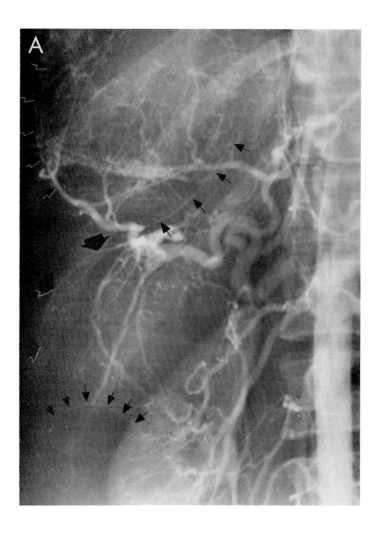

Comment: Liver cell adenoma is a rare hepatic neoplasm. Radiologic and pathologic differentiation between adenoma of the liver and hamartoma is frequently difficult. Adenomas of the liver are either liver cell or bile duct cell in type. A mixed cell type may be identical to the hepatic hamartoma. These tumors may be single or multifocal, as in the case presented here. In the presence of tumor vascularity, a malignant lesion must be considered; although no tumor blush occurred, there is evidence of arteriovenous shunting and puddling of contrast material, making consideration of hepatocellular carcinoma necessary.

Figure 75, courtesy of Palubinskas, A. J., Baldwin, J., and McCormack, K. R.: Radiology 89:444–445, September, 1967.

Figure 76.—Hepatic abscess.

A young man had received gunshot wounds in the right abdomen. Postoperatively, fever and signs of sepsis developed.

A, selective common hepatic angiogram, late arterial phase: The distal intrahepatic arterial branches, indicating the lateral extremity of the right lobe of the liver, are displaced medially by a mass between the diaphragm and the liver (**open arrows**). Two intrahepatic masses are indicated by draping and displacement of smaller intrahepatic branches. One is situated laterally and superiorly within the right lobe, and some moderately irregular arteries are visible in this region (**arrowheads**). The second mass is within the left lobe, partially overlying the lower dorsal vertebrae and displacing the branches of the left hepatic artery laterally to the right of the midline (**arrows**).

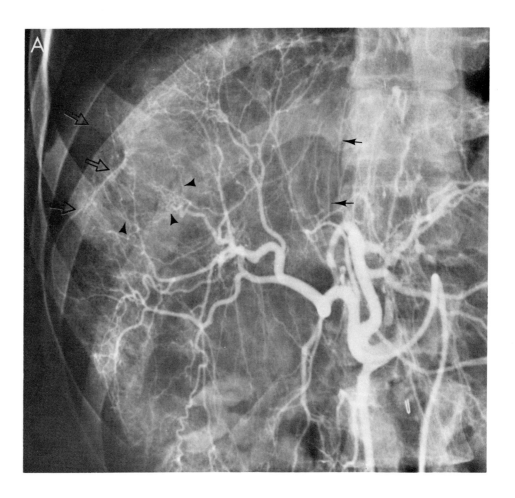

B, selective common hepatic arteriogram, hepatogram phase: The lateral outline of the right lobe of the liver is irregularly indented and displaced medially (**open arrows**). An irregular rim of contrast material surrounds a mottled-appearing mass in the right lobe (**arrows**). The left lobar mass seen in **A** is not demonstrated here. A hepatic vein is evident, and its smooth medial border may indicate displacement by the mass in the left lobe.

Comment: Most descriptions of the angiographic findings in intrahepatic abscess include the appearance of an irregular capsular blush in the late phase of the arteriogram. This commonly surrounds a lucent area which represents the abscess contents. The halolike blush, which surrounds the radiolucent center, may be the result of opacification of numerous vessels within inflammatory granulation tissue. Irregular small vessels resembling tumor vessels may be evident around the abscess. The arteriographic findings in this case also indicate an extrahepatic subdiaphragmatic mass that displaces and indents a lateral border of the right lobe and is evidence of a subdiaphragmatic abscess. Infection may spread to the contiguous liver from a subdiaphragmatic abscess. In this case, penetrating trauma had caused both the extrahepatic and the intrahepatic abscess.

Figure 73, courtesy of Dr. S. Baum, Massachusetts General Hospital, Boston.

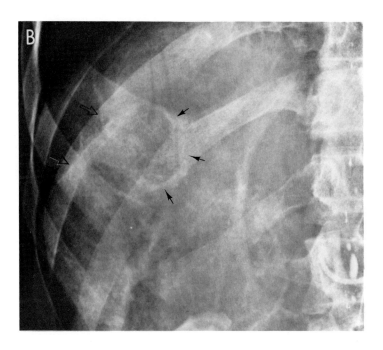

Figure 77.—Hepatic abscess.

A 28-year-old man had an extensive operation for repair of biliary stricture followed by the development of fever and signs of abdominal sepsis. Plain radiographs, apart from hepatic enlargement, are seldom revealing with intrahepatic abscess. In this case, communication of the abscess with the gastrointestinal tract resulted in gas shadows within the intrahepatic lesion. The angiographic pattern is bizarre. The central pool of contrast material within the lesion indicates some extravasation or bleeding within the abscess. The extensive tissue necrosis resulted in the lucent defect in the hepatogram and the mottled appearance of the surrounding parenchyma. The appearance of the branches of the right hepatic artery indicates that irregular vascularity is not evidence of neoplasm alone.

A, upper gastrointestinal radiograph, anteroposterior projection: Irregular small gas shadows are visible in the right upper abdomen in the region of the liver (**arrows**). An opacified fistulous tract communicates with the lesser curvature of the stomach and an inflammatory mass within the liver (**arrowheads**). The stomach is displaced inferiorly and laterally by the enlarged liver. The T-tube is seen with its long limb in the duodenum.

(Continued.)

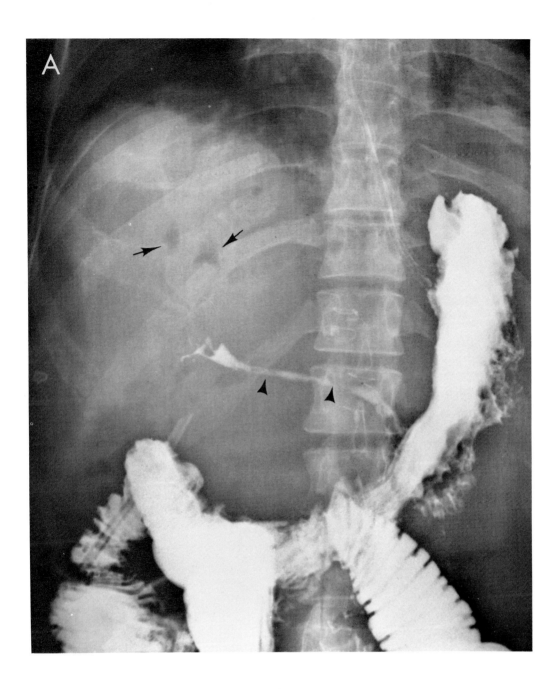

Figure 77 · Hepatic Abscess / 195

Figure 77 (cont.).—Hepatic abscess.

B, selective celiac angiogram, arterial phase: The right hepatic artery and the gastroduodenal artery are stretched by the large liver mass. The secondary branches of the right hepatic artery are markedly irregular (**arrowheads**). There is retrograde filling of an accessory left hepatic artery by way of the gastroepiploic and left gastric arteries.

C, selective celiac angiogram, hepatogram phase: A small pool of contrast material lies in a radiolucent defect within the hepatogram (**arrow**). The hepatic parenchyma contiguous to this defect is irregularly mottled.

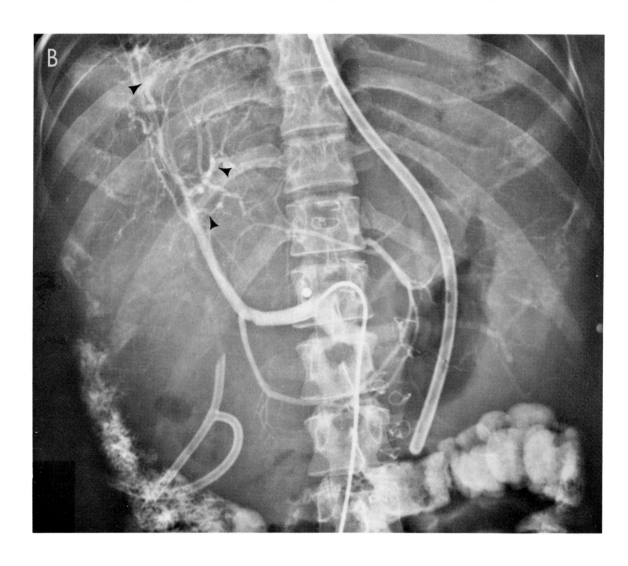

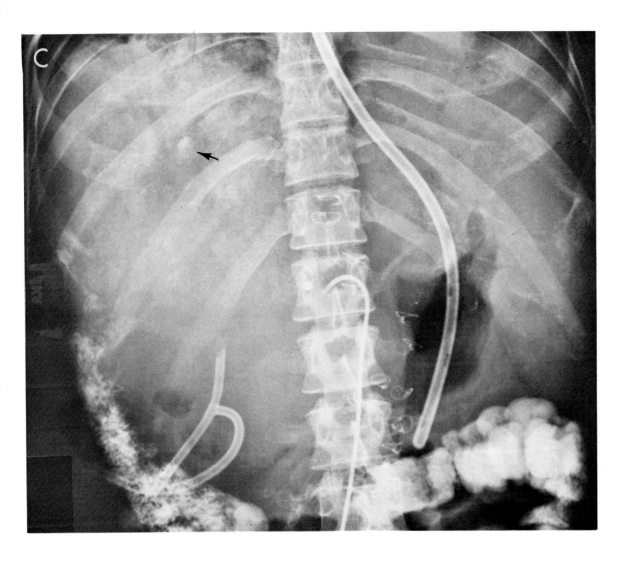

Figure 77 · Hepatic Abscess / 197

Figure 78.—Hepatic abscess.

A 52-year-old man complained of fever and right upper quadrant pain and tenderness eight weeks after a gastric resection and vagotomy. At operation, no subphrenic abscess was found.

A, selective celiac angiogram, arterial phase: There are marked stretching and separation of the distal branches of the right hepatic artery in the superior portion of the right lobe (**arrows**). A radiolucency on the plain radiograph of the abdomen before this study revealed gas in this region. The terminal branches of the hepatic artery (**arrowheads**) reach the periphery, indicating the site of the mass.

B, selective celiac arteriogram, capillary phase: Note a well-demarcated lucency within the right lobar hepatogram (**arrows**). An irregular rim of density surrounds it.

Comment: The classic signs of intrahepatic abscesses are demonstrated here, for example, arterial draping and displacement, a radiolucent defect and a surrounding rim of increased density.

Figure 78, courtesy of Fleming, R., *et al.*: Radiology 93:777, October, 1969.

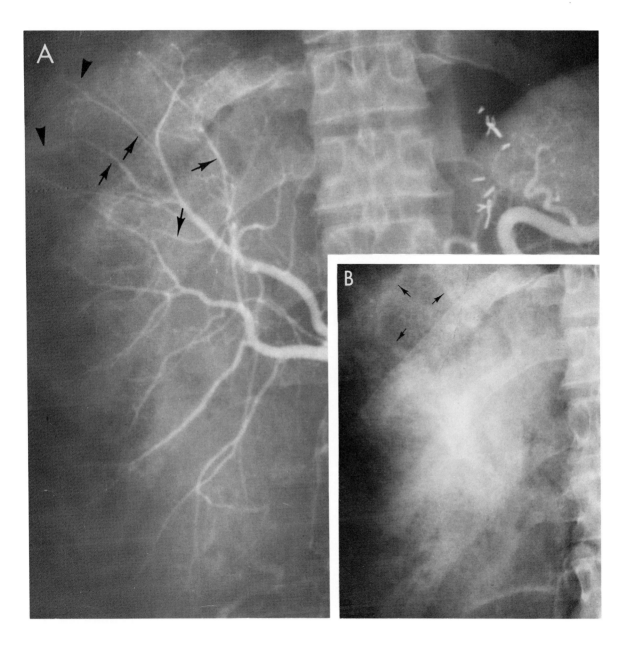

Figure 78 · Hepatic Abscess / 199

Figure 79.—Simple hepatic cyst.

A 64-year-old woman complained of anorexia, 10-lb. weight loss and right upper quadrant pain. Physical examination revealed the liver to be enlarged to 8 cm below the right costal margin.

A, upper gastrointestinal radiograph, anteroposterior projection: The stomach is displaced inferiorly and laterally (**arrowheads**) by a large hepatic mass.

B, selective celiac angiogram, arterial phase: There are marked stretching and displacement of the secondary branches of the right hepatic artery by a huge avascular mass either within or compressing the right lobe of the liver (**arrows**). The gastroduodenal artery and the proper hepatic artery are displaced medially.

(*Continued.*)

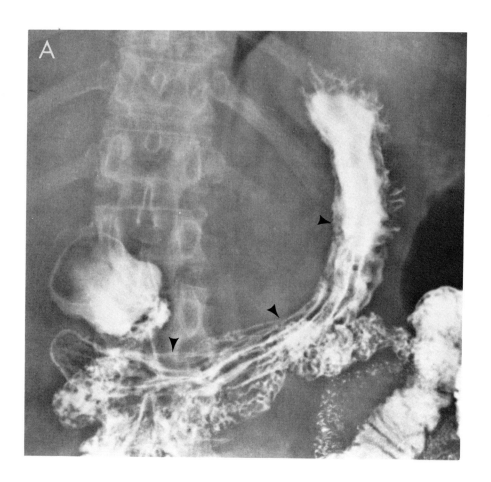

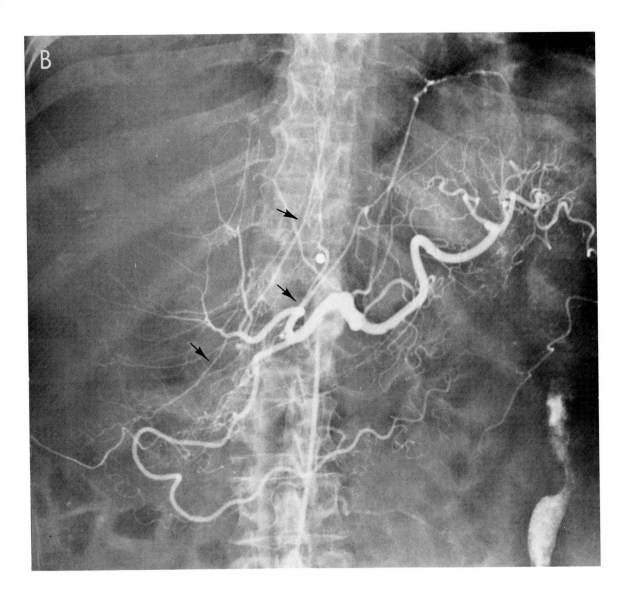

Figure 79 · Simple Cyst / 201

Figure 79 (cont.).—Simple hepatic cyst.

C, selective celiac angiogram, hepatogram phase: Some terminal branches of the right hepatic artery course around the right upper quadrant mass, suggesting that the mass is intrahepatic. Some compression of the liver parenchyma is noted medially. The tumor is avascular.

D, selective celiac angiogram, portal venous phase: The branch of the

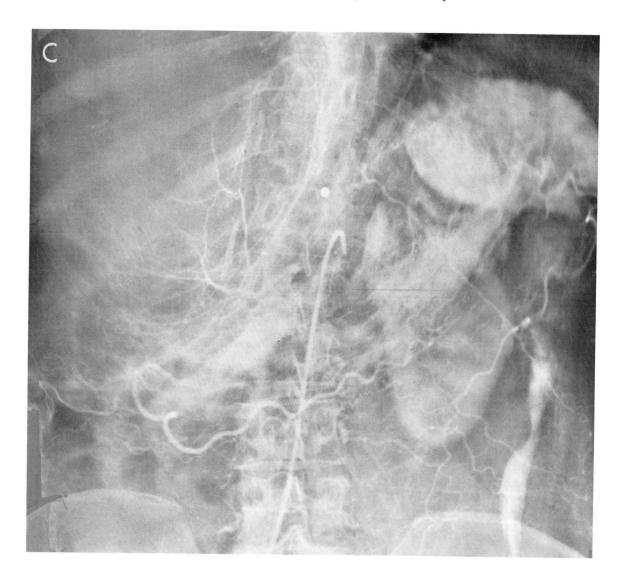

portal vein to the left lobe is well seen (**arrowheads**). The right portal vein is obstructed as a result of compression by the large right intrahepatic mass.

Comment: The etiology of simple cysts of the liver is varied and in many cases cannot be identified. Some are considered congenital, and a relationship to polycystic liver disease and polycystic kidney disease has been proposed. Some cysts of the liver are considered to be secondary to inflammatory conditions or trauma. All are characterized angiographically by displacement of intrahepatic vessels and complete absence of vascularity or blush.

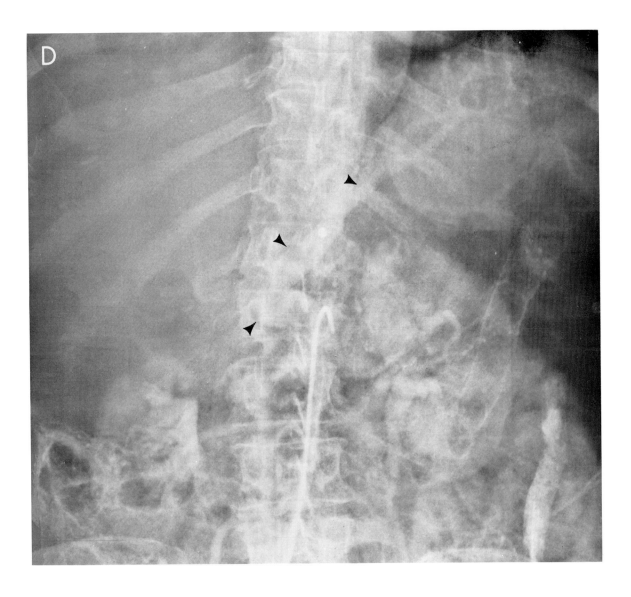

Figure 80.—Simple cyst of the liver.

A 33-year-old woman had a long history of chronic liver disease, with portal cirrhosis first diagnosed at 11 years of age. Hepatosplenomegaly was noted at age 14, and splenectomy was performed. The cyst of the liver was marsupialized at age 17. She had experienced abdominal pain for three years prior to this examination and had lost 15 lb. Tender swelling was palpable in the epigastrium separable from the liver. The liver was palpably enlarged.

A, upper gastrointestinal radiograph, right anterior oblique projection: The duodenal bulb and proximal duodenal loop show evidence of extrinsic pressure and are deviated laterally and anteriorly by a large mass (**arrows**).

B, intravenous cholangiogram, left anterior oblique tomographic projection: The common bile duct shows smooth displacement medially and posteriorly by a large liver mass (**arrowheads**). The common bile duct appears to be intrinsically normal.

(*Continued.*)

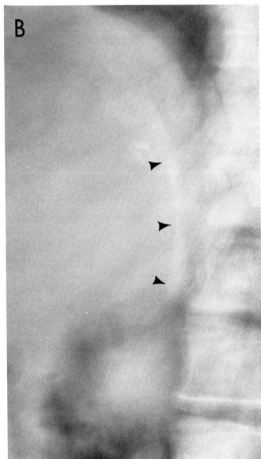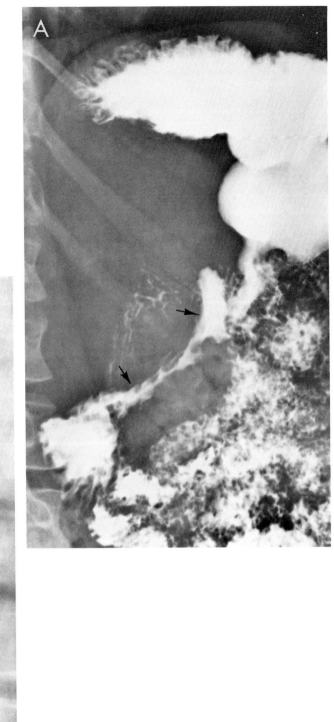

Figure 80 · Simple Cyst / 205

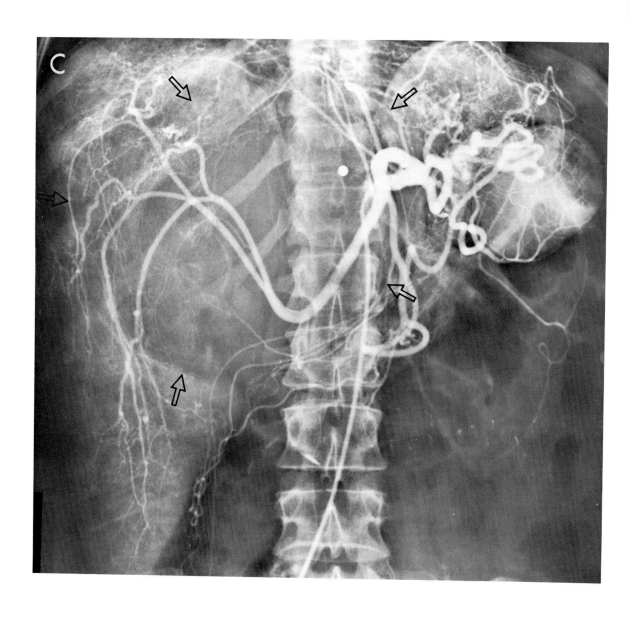

Figure 80 (cont.).—Simple cyst of the liver.

C, selective celiac angiogram, late arterial phase: A large avascular mass occupies most of the left lobe of the liver and a great part of the right lobe medially and superiorly (**arrows**). There is smooth margination of the mass. The hepatic arteries are displaced and draped over it. No tumor vasculature is evident.

D, selective celiac angiogram, hepatogram phase: The smoothly marginated large avascular mass is minimally lobulated. Again, no tumor vasculature or stain is evident.

(Continued.)

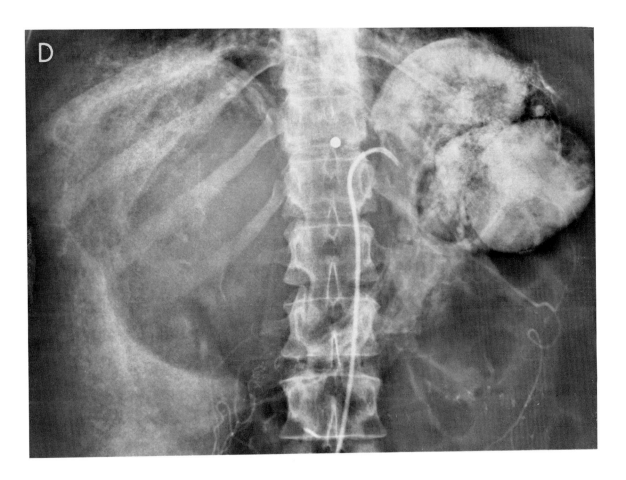

Figure 80 (cont.).—Simple cyst of the liver.

E, liver scan, anteroposterior projection: The liver is enlarged. Note a large area of deficient uptake of the radionuclide in the medial portion of the right lobe and most of the left lobe of the liver. The appearance conforms to the angiographic findings.

F, liver scan, right lateral projection: The region of deficient uptake lies anteriorly in the right lobe.

Comment: This case illustrates a type of simple cyst of the liver secondary to inflammatory or obstructive disease of the liver; in this instance, portal cirrhosis. Cysts are believed to arise in obstructed and dilated biliary ducts. They may be single or multiple.

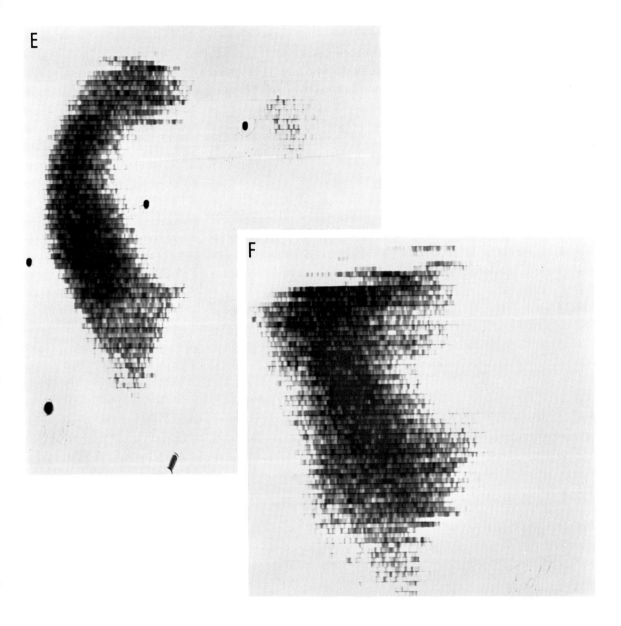

Figure 80 · Simple Cyst / 209

Figure 81.—Simple hepatic cyst.

A 69-year-old woman had had abdominal surgery, at which time multiple superficial cysts of the liver were noted. Aspiration revealed clear fluid.

A, selective celiac angiogram, late arterial phase: Note displacement of the terminal branches of the right hepatic artery in two areas, one superiorly and one laterally (**arrowheads**). The hepatic arteries in general are mildly tortuous. No tumor vasculature is seen.

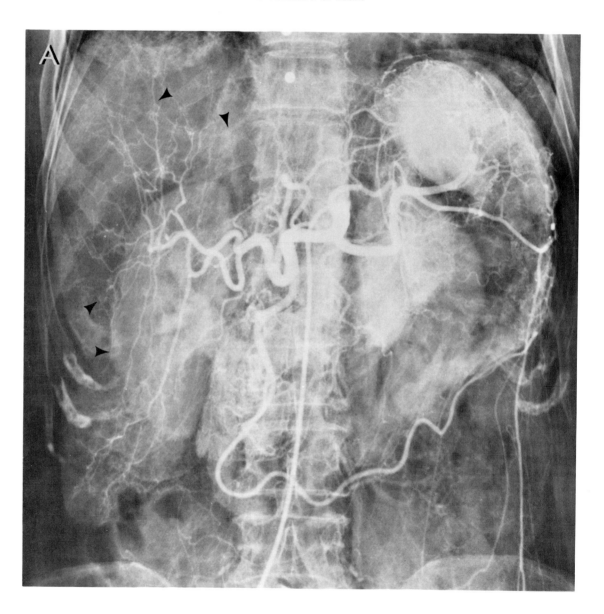

B, selective celiac angiogram, late hepatogram phase: Two radiolucent defects (**arrowheads**) in the hepatogram correspond to the regions of vessel displacement seen in **A**. These defects are smoothly outlined and completely avascular.

Comment: The generalized tortuosity of the smaller branches of the hepatic arteries in this case is a pattern most often associated with cirrhosis, but this diagnosis was not confirmed here.

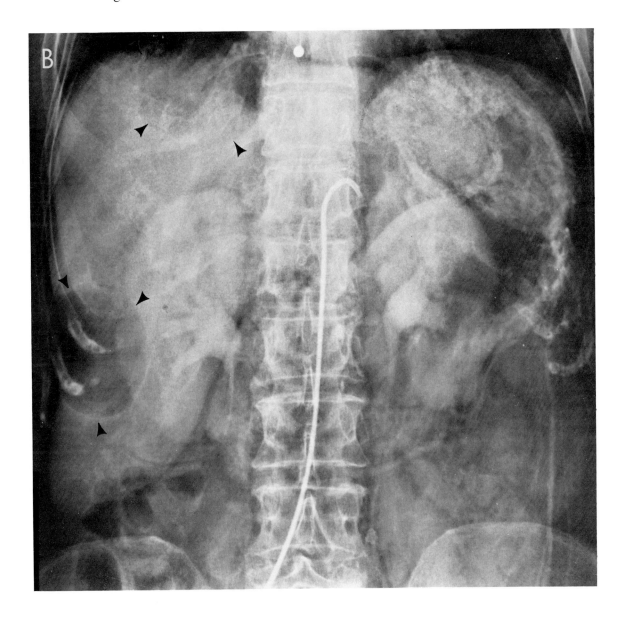

Figure 82.—Hydatid cysts of the liver.

Abdominal radiograph, anteroposterior projection: Crescentic circular and oval rimlike calcific densities are present in the right upper quadrant of the abdomen.

Comment: These calcifications are characteristic of hydatid cyst. Calcification in the capsule of hydatid cysts is common in the liver but not in the lung. The smaller calcifications are in daughter cysts. Other right upper abdominal calcifications that might be confused include those in the wall of renal cysts, in adrenal tumors, in laminated biliary calculi and in porcelain gallbladder.

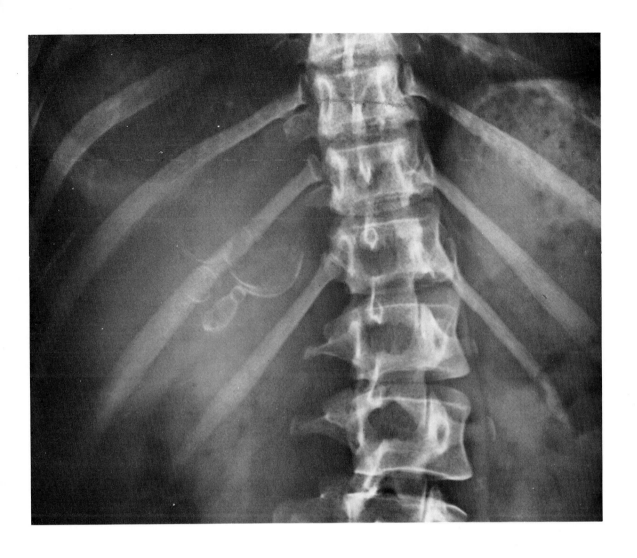

Figure 82 · Hydatid Cysts / 213

Figure 83.—Hydatid cyst of the liver.

A 48-year-old man complained of sharp epigastric pain while in a body cast after operation for spinal fusion. After removal of the cast, an upper abdominal mass was palpated.

A, selective celiac angiogram, late arterial phase: The left gastric artery is elevated and the splenic artery is depressed. There is marked separation of the branches of the left hepatic artery (**arrows**) encompassing a large mass within the left lobe of the liver. The right hepatic artery and its branches are displaced laterally. **Lh** = left hepatic artery; **G** = left gastric artery; **S** = splenic artery.

B, selective celiac angiogram, portal venous phase: The splenic vein is depressed. The inferior margin of the mass is delineated by a rim of density (**arrows**). The mass is completely avascular and radiolucent in comparison to the parenchyma of the liver. **SV** = splenic vein.

Comment: The dense rim surrounding the radiolucent mass demonstrated in this study is a frequent finding on angiography of hepatic hydatid cyst. The hydatid cyst has three layers; the outer, the ectocyst, is composed of compressed and scarred host tissue. This scar tissue is vascular, and it is postulated that this vascularized scar tissue is responsible for the rim of increased density.

Figure 83, courtesy of Fleming, R., *et al.*: Radiology 93:774, October, 1969.

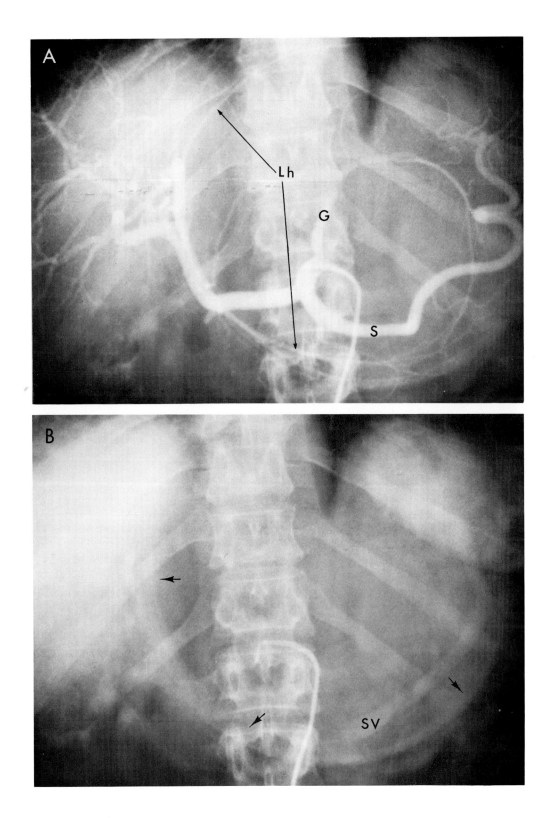

Figure 83 · Hydatid Cyst

Figure 84.—Hydatid cysts of the liver.

A, chest radiograph, posteroanterior projection: A smooth well-defined mass impinges on the right lower lobe. The lower border of the mass cannot be defined. The lateral portion of the right diaphragmatic leaf is not seen below the lesion. There appears to be pleural thickening and possibly some minimal pleural fluid at the right costophrenic angle.

B, chest radiograph, lateral projection: The mass is inseparable from the right diaphragmatic leaf and contiguous with the diaphragmatic shadow. Pleural thickening is again visible lying anterior to the mass.

C, abdominal radiograph, posteroanterior projection: The mass appears to be of uniform density and is inseparable from the right hemidiaphragm and the liver. The transverse linear shadow through the mass does not represent an air-fluid level; it is the result of pleural thickening.

Comment: Hydatid cysts commonly occur in the superior portion of the right lobe of the liver. Growth here may cause a localized elevation of the diaphragm and the mass then impinges on the thoracic cavity. Some difficulty is experienced in differentiating chest lesions from diaphragmatic or subdiaphragmatic tumors in these instances. The pleural thickening in this case might be mistaken for an air-fluid level in a cavity. Intrathoracic rupture of hydatid cysts is not infrequent and results in formation of pulmonary hydatid cysts. Basilar atelectasis and pleural effusion and thickening are frequent coincidental findings.

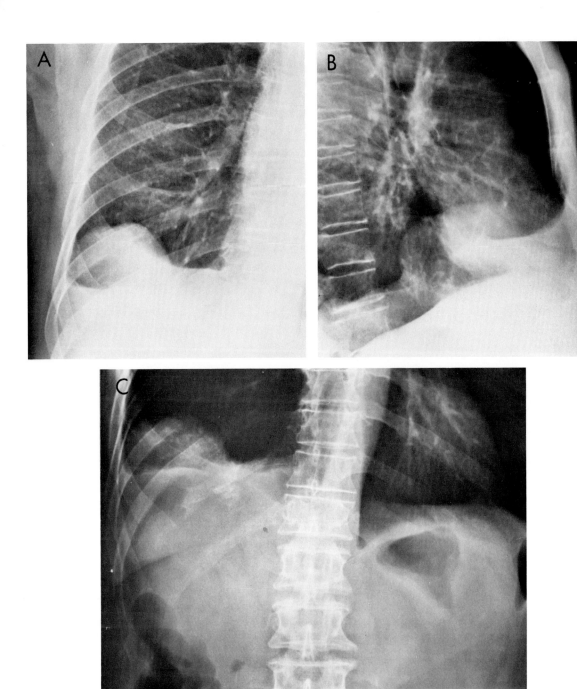

Figure 84 · Hydatid Cysts

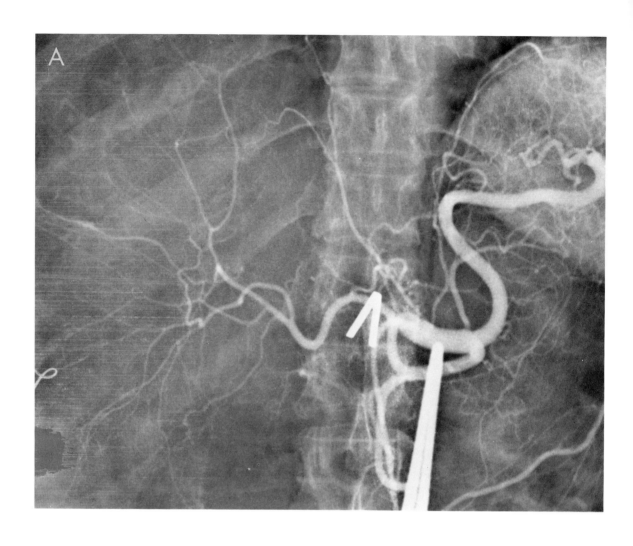

Figure 85.—Polycystic disease of the liver.

A 60-year-old woman had had bilateral mastectomy and bilateral oophorectomy for breast cancer. Nine years later she was hospitalized for investigation of hepatosplenomegaly with a presumptive diagnosis of metastatic liver disease. The liver was enlarged 6 cm below the right costal margin, and the spleen was palpable. Results of liver function tests were normal. Polycystic liver and kidney disease were confirmed surgically. No metastases were found in the liver.

A, selective celiac angiogram, arterial phase: The secondary branches of the hepatic artery within the liver are thin and attenuated and appear to be minimally separated. Terminal branches at the periphery do not fill.

B, selective celiac angiogram, portal venous phase: The branches of the right portal vein are displaced and deformed (**arrows**). Multiple lucencies of various sizes are present in the portal hepatogram.

(*Continued.*)

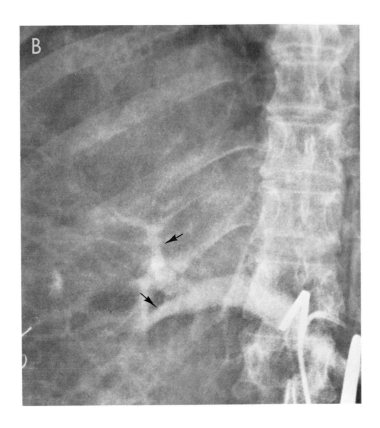

Figure 85 (cont.).—Polycystic disease of the liver.

C, right renal angiogram, capillary phase: The kidney is deformed and enlarged, and multiple lucencies are seen within the parenchyma.

Comment: The appearance of the right renal angiogram is diagnostic of polycystic kidney disease. The multiple lucencies within the portal hepatogram and the stretching and displacement of the arteries of the right lobe could be attributed to many diseases that cause similar angiographic findings, especially metastatic disease. However, in the light of the selective renal angiographic findings, polycystic disease of the liver must be suspected. Polycystic kidney disease will be found in association with multiple cysts in the liver in 15% of patients.

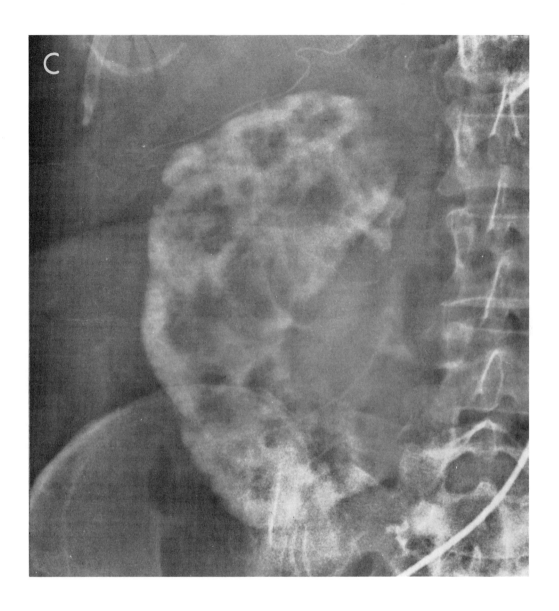

Figure 85 · Polycystic Disease of Liver / 221

Figure 86.—Diffuse chronic granulomas of the liver.

A young woman was seen because of fever and fatigue. Results of liver function tests proved abnormal and the patient's contraceptive pill was suspected as being the cause. One month later she was readmitted in a semicomatose condition. Blood cultures indicated a coagulase-positive streptococcus. Biopsy of the liver showed chronic inflammatory changes. A laparotomy, at a later date, confirmed the presence of multiple granulomas throughout the liver.

A, abdominal radiograph, anteroposterior projection: The stomach gas shadow is displaced laterally (**arrowheads**) and there is smooth effacement of the lesser curvature aspect of the stomach caused by left lobar hepatic enlargement.

B, selective celiac angiogram, arterial phase: The liver is enlarged, and the hepatic artery and its branches are greatly increased in caliber. The smaller intrahepatic arterial branches are irregular. A mottled vascular blush is seen throughout most of the liver. A few small irregularly avascular segments are present in the right lobe.

C, selective celiac angiogram, hepatogram phase: Enlargement of the liver is apparent. The early hepatogram shows a mottled appearance with some vascular and some avascular regions.

D, superior mesenteric angiogram, arterial phase: The hepatic artery has filled by retrograde flow in the pancreatic arcade and gastroduodenal artery. This indicates increased flow demand in the hepatic vascular bed.

Comment: Infection in the liver is usually manifested as a localized abscess, with well-circumscribed lucent defects being apparent on the hepatic angiogram. The appearance in this case is most unusual. The increase in size of the hepatic vessels plus the marked diffuse blush could be interpreted as hepatoma. A diagnosis of hemangioma or vascular metastasis could also be entertained.

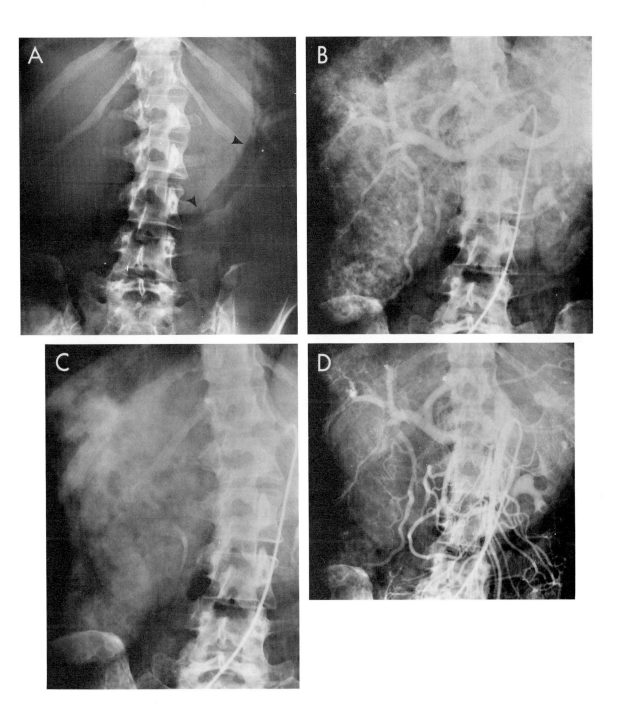

Figure 86 · Diffuse Chronic Granuloma

Figure 87.—Regenerated liver nodule.

A, upper gastrointestinal radiograph, anteroposterior projection, obtained 1 hr after ingestion of barium: Note smooth displacement of small bowel loops inferiorly and medially from the right upper abdomen (**arrows**). The stomach, duodenum, kidney and colon are not displaced.

B, superior mesenteric angiogram, arterial phase: The right hepatic artery has its origin from the superior mesenteric artery. The right lobe of the liver is greatly enlarged and the inferior margin extends to the right lower abdomen, corresponding to the pressure defect on the small bowel seen in **A**. The branches of the right hepatic artery in this region appear to be stretched and attenuated. The arterial branches in the right upper lobe show evidence of corkscrewing (**arrows**).

C, selective celiac angiogram, arterial phase: A small accessory right hepatic artery has its origin from the proper hepatic artery. The branches of this vessel are markedly irregular (**arrow**). The left lobe of the liver is greatly enlarged and the left hepatic artery and its branches are large.

D, selective celiac angiogram, portal venous phase: The right branches of the portal vein in the right lobe of the liver (**arrows**) are enlarged and extend inferiorly within the regenerated liver nodule.

Comment: The liver, in the process of repairing damage due to cirrhosis, may form nodular masses of regenerated parenchyma. The nodules may appear to be mass lesions within the liver and may mistakenly be interpreted as neoplasms. These regenerated nodules have a wide spectrum of angiographic presentation. The liver scan may differentiate between hepatoma and a regenerating liver nodule, as the nodule appears as an area of increased uptake on the radioactive gold liver scan. Hemangioma and vascular metastases must also be differentiated from regenerative nodules.

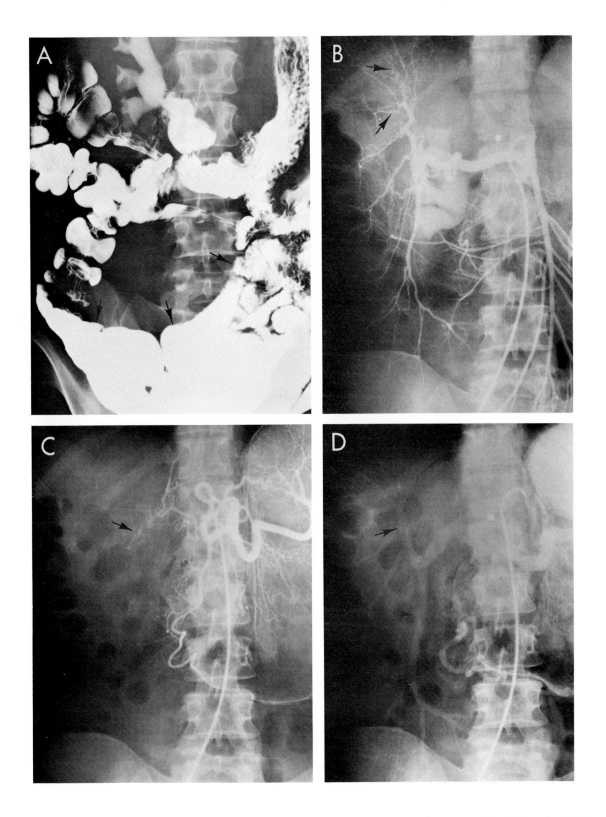

Figure 87 · Regenerated Liver Nodule / 225

PART 3

The Extrahepatic Bile Ducts

Characteristics of Tumors of the Extrahepatic Bile Ducts

Primary Malignant Tumors

Primary carcinoma of the bile ducts is an uncommon although not rare tumor. Its preoperative diagnosis is becoming more frequent, partly because of more sophisticated radiologic techniques and partly because of increased clinical awareness. The frequency of biliary duct carcinoma is approximately one-eighth that of pancreatic carcinoma. It comprises approximately 3% of neoplasms at autopsy, and it is found about once in every 300 biliary tract operations.

Carcinoma of the bile ducts is found more frequently in men than in women in a ratio of approximately 3:2, with the most common age of occurrence being in the sixth, seventh and eighth decades. Gallstones are coincident in 20% of cases. Controversy exists as to whether gallstones result from biliary stasis as a result of tumor or are a predisposing factor.

Carcinoma of the bile ducts occurs most often in the lower biliary tract, that is, the distal common bile duct and the ampulla of Vater. The second most frequent site of occurrence is in the proximal hepatic duct near the junction of the right and left hepatic ducts. The cystic duct and the right or left hepatic duct may be involved alone or in conjunction with neighboring ducts. Often, the exact site of origin is doubtful because of the propensity for this tumor to extend intramurally along the biliary tract. As a result of its capability for extension, the tumor quite commonly has infiltrated widely before symptoms are noted. It has been claimed that these tumors may arise in papillomas of the biliary tree. However, in view of the extreme rarity of papillomas of the biliary tree and the relative frequency of biliary duct carcinoma, it is difficult to envision many carcinomas arising in biliary duct papillomas.

Three major anatomic types are described. The least frequent is a papillary tumor that grows intraluminally, similar to polypoid carcinomas in the gastrointestinal tract. The tumor may fill and obstruct the lumen and may be multicentric in origin. Two other gross pathologic types are about equal in frequency. One is the annular infiltrating type which causes a short stricture, commonly with overhanging margins, and the other is the scirrhous type. The latter usually involves a single site, but many areas may be involved in an irregular pattern similar to sclerosing cholangitis. It is a dense, white fibrous type of tumor which extends largely intramurally. The mucosa overlying these tumors may appear grossly to be normal. The strictures caused by these scirrhous types of tumors are smoothly tapering.

The majority of adenocarcinomas of the bile ducts are columnar cell in type, with occasional cases of squamous cell carcinoma, adenoacanthoma and leiomyosarcoma being found.

Metastases are found at operation in one-third of the cases. Metastasis occurs early in the polypoid and infiltrating types and late in the scirrhous type. Metastasis may be local, as a result of direct extension, distant by way of the lymphatic system or may be blood-borne. These latter metastases commonly involve the regional nodes, the liver and the lungs. The tumor may extend along the bile ducts to the liver and gallbladder. The pain of bile duct carcinoma may be the result of its propensity for nerve invasion that occurs in 63% of tumors. Despite the proximity of the portal vein, invasion of this structure rarely occurs. The peritoneum may be involved by local extension. In these cases ascites frequently occurs.

CLINICAL FEATURES.—The poor prognosis of biliary duct carcinoma is due in part to its lack of symptomatology in early stages. About 20% of these tumors are correctly diagnosed clinically. The proximity of many vital structures and diseases which cause similar symptoms makes definitive diagnosis difficult.

Jaundice is the most common clinical finding, present with 90% of biliary duct carcinomas at the time of diagnosis. It may be insidious in onset or may be abrupt. In 60% of cases it is fluctuating. Pain, consistent in type rather than colicky, is the most frequent accompanying symptom. Weight loss, cachexia, pruritus, anorexia, fever, chills, dyspepsia and vomiting are other frequent complaints. In many instances weight loss will be noted before jaundice but may not bring the patient to the physician. If the tumor is located in the cystic duct alone, jaundice may not occur. Jaundice may develop despite an incompletely obstructing tumor. Biliary stasis may occur in the absence of complete obstruction because of intramural infiltration of the tumor and decreased ductal pliability. The gallbladder is palpable in one-third to one-half of patients. A tumor must, of course, occur below the level of the cystic duct, and the gallbladder itself must be free of disease for this to occur. The liver is palpably enlarged in 70–80% of patients. The results of liver function tests are consistent with obstructive jaundice. However, prolonged obstruction of the biliary tree results in hydrohepatosis and liver cell damage, which may be reflected in a change in the liver function test pattern. Carcinoma of the bile duct is occasionally complicated by suppurative cholangitis or empyema of the gallbladder.

RADIOLOGIC FEATURES.—A conventional radiograph of the abdomen most frequently shows liver enlargement and little else. The upper gastrointestinal examination may show changes secondary to tumors that involve the lower biliary tract, for example, the distal common bile duct or the ampulla of Vater. The most frequent finding on the upper gastrointestinal examination is an extrinsic pressure defect on the proximal duodenum. A nodular intraluminal lesion projecting from the medial wall of the duodenum is often evident with ampullary carcinomas. Greater accuracy of diagnosis compared with the conventional examination is supplied by hypotonic duodenography in cases of distal biliary carcinomas. These are indirect methods of examination and the findings are similar to those occurring with other causes of lower biliary obstruction such as an impacted calculus in the distal common bile duct, acute or chronic pancreatitis and carcinoma of the head of the pancreas. Intravenous cholangiography, because of failure to delineate the ducts in jaundiced patients, is usually not indicated. There is little use in attempting study in patients with a serum bilirubin value more than 4 mg/100 ml and in those with a sulfobromophthalein retention value more than 40%. As most patients with carcinoma of the biliary tract are deeply jaundiced, nonvisualization is the rule. Percutaneous transhepatic cholangiography is the radiologic examination of choice when biliary duct carcinoma is suspected. In many cases, at the time of examination, a question still exists as to whether jaundice is due to parenchymal disease or to obstruction. Opacification of dilated biliary ducts immediately answers this question. Percutaneous cholangiography gives direct evidence of the site and character of the obstructing lesion. The character of the lesion is deduced from the shape of the termination of the contrast column within the bile duct. When contrast material passes beyond the lesion, the size and shape of the lesion itself is defined. Lesions that must be differentiated on the percutaneous cholangiogram are obstruction as a result of impacted calculi, chronic or acute pancreatitis, pancreatic carcinoma and extrinsic pressure due to hilar metastasis. The ducts proximal to the lesion in biliary duct carcinoma are often distensible and may reach an extremely large size.

The angiographic findings are quite similar to those described for carcinoma of the gallbladder. The most frequent finding is encasement of arteries in the region of the hilus of the liver. The right hepatic artery is most commonly involved by this process. The portal vein is obstructed in relatively few instances; this occurrence is unexplained. Little, if any, tumor vascularity or blush is noted.

A new dimension has been added to the diagnosis of these lesions with the advent of fiberoptic endoscopy coupled with direct cannulation of the biliary tract. This promises to eliminate, in large part, the use of percutaneous transhepatic cholangiography.

BENIGN TUMORS

Benign tumors of the extrahepatic bile ducts are extremely rare. Malignant tumors are 10 times as frequent as benign tumors. The intrahepatic ducts are a more frequent site of benign tumors than are the extrahepatic ducts. Of the benign tumors, the papilloma is the most common. The common bile duct is frequently the site of origin. It may cause obstruction and occasionally there is dilatation of the bile ducts proximal to the tumor. These tumors are more often sessile than pedunculated. They are made up of numerous papillae covered by columnar epithelium with central connective tissue support. Adenomas are less frequent. Fibromas have been known to occur in any position along the extrahepatic biliary tract. Postcholecystectomy amputation neuromas may rarely occur at the site of cystic duct amputation. They are composed of nerve axons and scar tissue. A small, firm, bulbous mass is found imbedded in the blind end of the duct. Chronic severe pain occurs with this tumor. It follows cholecystectomy by an interval of three to six months. Other benign tumors that have been described as occurring in the extrahepatic biliary ducts are lipoma and leiomyoma.

CHOLEDOCHAL CYST.—The choledochal cyst or choledochocele is an idiopathic dilatation of the common bile duct. Most are believed to be congenital. They are not true cysts but localized dilatations or enlargements of the common bile duct.

They most commonly involve the supraduodenal portion of the common bile duct but may be as proximal as the common hepatic duct. The dilatation is circumferential in most cases, and the cysts vary greatly in size. The cystic wall is commonly thick and contains fibrous tissue. The mucosal lining may be denuded.

Choledochal cysts affect women more commonly than men. They frequently are manifested in childhood or early adult life. Presenting complaints are usually abdominal pain and jaundice. A right upper quadrant abdominal mass is palpable in 77% of cases. They usually run a chronic intermittent course.

Intravenous cholangiography and the upper gastrointestinal series are useful radiographic examinations in the diagnosis of choledochal cyst. Direct cannulization will undoubtedly prove useful also. The cyst may opacify on intravenous cholangiography, and a dilated and partially obstructed common bile duct may be seen. There may be extrinsic pressure on the duodenum and this will be apparent on the upper gastrointestinal series.

SUGGESTED READING

Braasch, J. W., Warren, K. W., and Kune, G. A.: Malignant neoplasms of the bile ducts, Surg. Clin. North America 47:627–638, June, 1967.

Bree, R. L., and Flynn, R. E.: Hypotonic duodenography in the evaluation of choledocholithiasis and obstructive jaundice, Am. J. Roentgenol. 116:309–319, October, 1972.

Clemett, A. R.: Carcinoma of the major bile ducts, Radiology 84:894–903, May, 1965.

Cohn, E. M.: Tumors of the Gallbladder and Bile Ducts, in Bockus, H. L. (ed.): *Gastroenterology* (2nd ed.; Philadelphia: W. B. Saunders Company, 1965), Vol. 3, Chap. 16, pp. 811–26.

Eaton, S. B., et al.: Hypotonic duodenography, Radiol. Clin. North America 8:125–137, April, 1970.

Fleming, M. P., Carlson, H. C., and Adson, M. A.: Percutaneous transhepatic cholangiography: The differential diagnosis of bile duct pathology, Am. J. Roentgenol. 116:327–336, October, 1972.

Kaude, J., and Rian, R.: Cholangiocarcinoma, Radiology 100:573–580, September, 1971.

Krieger, J., Seaman, W. B., and Porter, M. R.: The roentgenologic appearance of sclerosing cholangitis, Radiology 95:369–375, May, 1970.

Legge, D. A., and Carlson, H. C.: Cholangiographic appearance of primary carcinoma of the bile ducts, Radiology 102:259–266, February, 1972.

Mujahed, Z., and Evans, J. A.: Percutaneous transhepatic cholangiography, Radiol. Clin. North America 4:535–545, December, 1966.

Mujahed, Z., and Evans, J. A.: Pseudocalculus defect in cholangiography, Am. J. Roentgenol. 116:337–341, October, 1972.

Okuda, K., et al.: Endoscopic pancreaticocholangiography, Am. J. Roentgenol. 117:437–445, February, 1973.

Wise, R. E., and O'Brien, R. G.: Interpretation of the intravenous cholangiogram, J.A.M.A. 160:819–827, Mar. 10, 1956.

Figure 88.—Carcinoma of the common bile duct.

A 58-year-old man had upper abdominal pain for four weeks before the onset of jaundice that had been present for two weeks before admission to another hospital. At laparotomy, a T-tube was inserted into the common bile duct, and he was referred to the Lahey Clinic for further evaluation.

T-tube cholangiogram, anteroposterior projection: The intrahepatic and extrahepatic biliary ducts are dilated proximal to an obstructing lesion in the common bile duct. There is gross dilatation of the common bile duct immediately proximal to the obstruction. The termination of the column of contrast material shows slight irregularity of outline, that is, an irregular concave meniscus (**arrows**). No contrast material passes beyond the obstruction.

Comment: The upwardly concave meniscus is more characteristic of carcinoma of the pancreas than of primary carcinoma of the common bile duct. The irregularity of the termination of the common bile duct in this region, however, suggests an intrinsic tumor. The bile ducts proximal to the malignant lesion are capable of extensive dilatation.

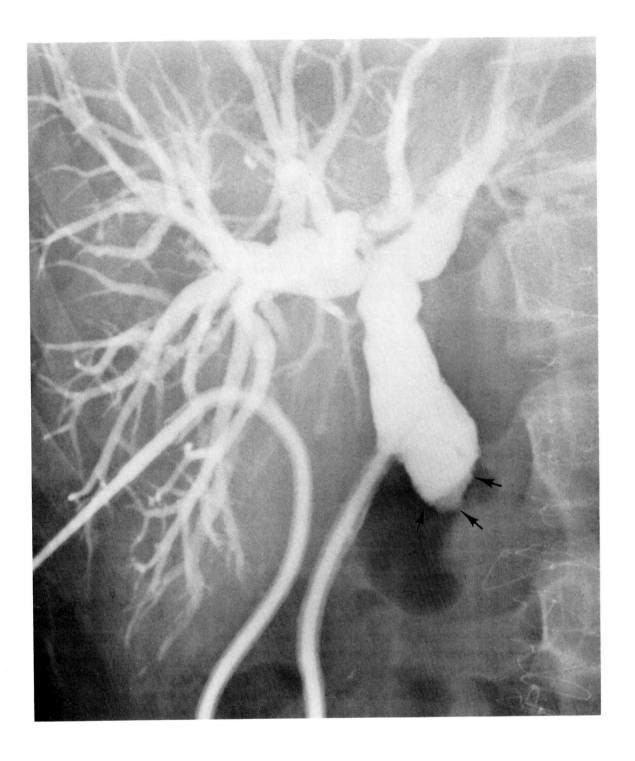

Figure 88 · Carcinoma of Common Bile Ducts / 235

Figure 89.—Carcinoma of the common bile duct.

An 82-year-old man complained of acute abdominal pain. He was a poor surgical risk and was not operated on; therefore, surgical or pathologic proof is not available.

A, intravenous cholangiogram, anteroposterior tomogram: There is marked dilatation of the common bile duct, common hepatic duct and the intrahepatic ducts. The common bile duct is obstructed distally. The appearance of the common bile duct at the point of obstruction is irregularly concave (**arrows**).

B, percutaneous transhepatic cholangiogram, spot film, anteroposterior projection: The bile ducts are grossly dilated proximal to an annular constricting lesion in the common bile duct. The proximal border of this is well defined and has overhanging margins (**arrows**). A small amount of contrast medium passes beyond the tumor and outlines the napkin-ring type of lesion.

Comment: When the shape and character of the lesions are clearly defined as in the percutaneous cholangiogram, the radiologic examination is diagnostic. Intravenous cholangiography is indicated in very few instances because of the poor chance of opacification in jaundiced patients. In this case, even though the site of obstruction is localized on the intravenous cholangiogram, the character of the obstructing lesion is not sufficiently defined for precise diagnosis.

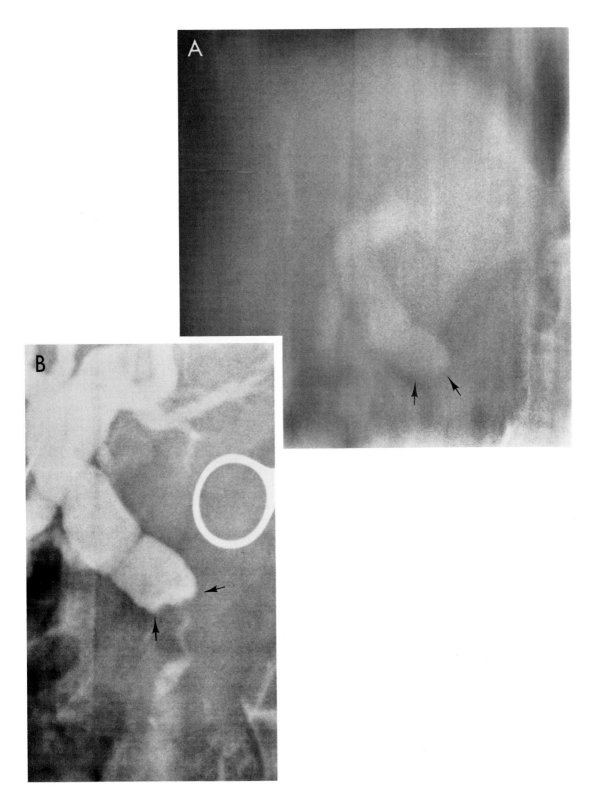

Figure 89 · Carcinoma of Common Bile Ducts / 237

Figure 90.—Carcinoma of the common bile duct.

A 48-year-old man with a previous history of pancreatitis had a recent onset of painless jaundice.

A, T-tube cholangiogram, anteroposterior projection: The intrahepatic ducts, the common hepatic duct and the common bile ducts are dilated. The distal common bile duct is tortuous, with marked irregularity and nodularity of the mucosa (**arrows**).

B and **C,** T-tube cholangiogram, spot films in different projections: The irregularity and nodularity of the distal common bile duct are clearly demonstrated (**arrows**).

Comment: This is an extremely rare presentation of carcinoma of the ampulla and distal common bile duct. There is extensive dilatation of the proximal biliary ducts despite the lack of complete obstruction. Tumor invasion along the wall of the ducts contributes to biliary stasis and development of jaundice. In this case the tumor involves a long segment of the distal common bile duct as shown by the irregular nodular appearance.

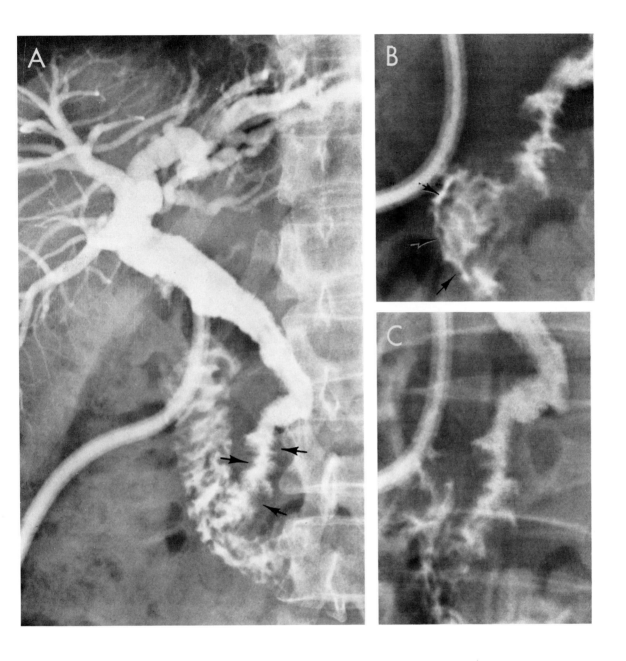

Figure 90 · Carcinoma of Common Bile Ducts / 239

Figure 91.—Carcinoma of the common bile duct.

This 78-year-old woman was seen with recurrent jaundice two months after a cholecystectomy and common bile duct exploration. She also complained of decreased appetite and weight loss. The liver was tender and palpable 2 cm below the right costal margin.

Percutaneous transhepatic cholangiogram, anteroposterior projection: The right hepatic biliary ducts are grossly dilated, as are the common hepatic duct and the proximal common bile duct. The left hepatic duct and its branches are not demonstrated. A small quantity of contrast material passes distal to the point of obstruction, but the structures in this area are ill-defined.

Comment: Nonfilling of a branch of the biliary tree is an extremely important finding during percutaneous transhepatic cholangiography. If enough care has been taken in attempting to fill the entire biliary tree, such a finding indicates obstruction. In this case, carcinoma of the common bile duct had extended proximally to involve the common hepatic duct and the left hepatic duct.

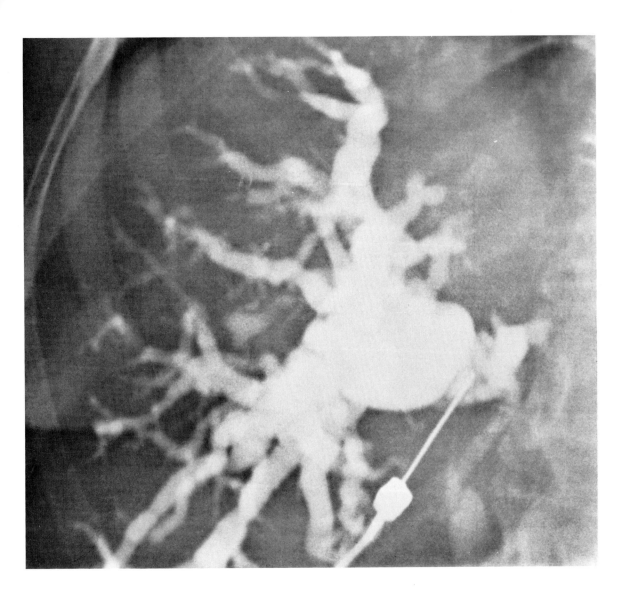

Figure 91 · Carcinoma of Common Bile Ducts / 241

Figure 92.—Carcinoma of the bile ducts.

Percutaneous transhepatic cholangiogram, anteroposterior projection: Numerous areas of narrowing and dilatation are diffusely scattered throughout the intrahepatic biliary tree. Apart from mild dilatation, the extrahepatic bile ducts appear to be normal.

Comment: Carcinoma of the bile duct may diffusely involve the biliary tract and cause segmental narrowing, as in this case. This appearance may be confused with sclerosing cholangitis. The extrahepatic bile ducts are almost always involved in sclerosing cholangitis. Absence of marked abnormality of the extrahepatic ducts in this case is against the diagnosis of sclerosing cholangitis. A small group of bile duct cancers are segmental in nature, grow slowly and give a clinical and radiologic presentation similar to that of sclerosing cholangitis.

Figure 92, courtesy of Drs. R. J. Cobb and E. J. Sennett, St. Francis Hospital, Hartford, Conn.

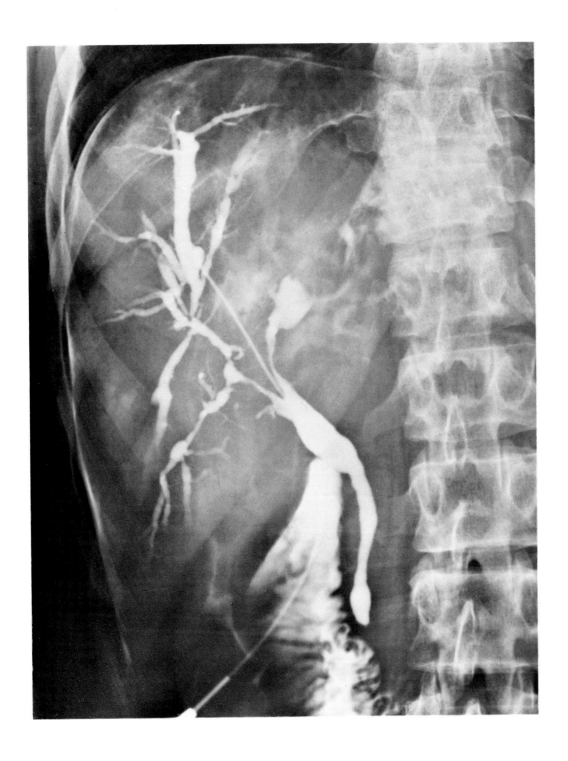

Figure 92 · Carcinoma of the Bile Ducts / 243

Figure 93.—Carcinoma of the common hepatic duct.

A 58-year-old woman was admitted to another hospital because of abdominal pain and recent onset of jaundice. At laparotomy, a choledochotomy revealed papillary adenocarcinoma of the bile duct and metastasis to the liver.

A, T-tube cholangiogram, left anterior oblique projection: Note napkin-ring type of lesion partially obstructing the proximal common hepatic duct. Contrast material outlines the overhanging margins of the tumor. The branches of the right hepatic duct are more moderately dilated. The left hepatic duct, being completely obstructed, is not visible.

B, T-tube cholangiogram, spot film, lateral projection: The sharply overhanging margins of the tumor are clearly demonstrated (**arrows**). The appearance is similar to that of the annular type of carcinoma of the colon. This projection also demonstrates nonfilling of the left hepatic ducts.

Comment: In our experience, this annular medullary type of lesion occurs with about equal frequency to the sclerosing type with a smoothly tapered segment of narrowing. The segment of involvement is commonly short in both pathologic types.

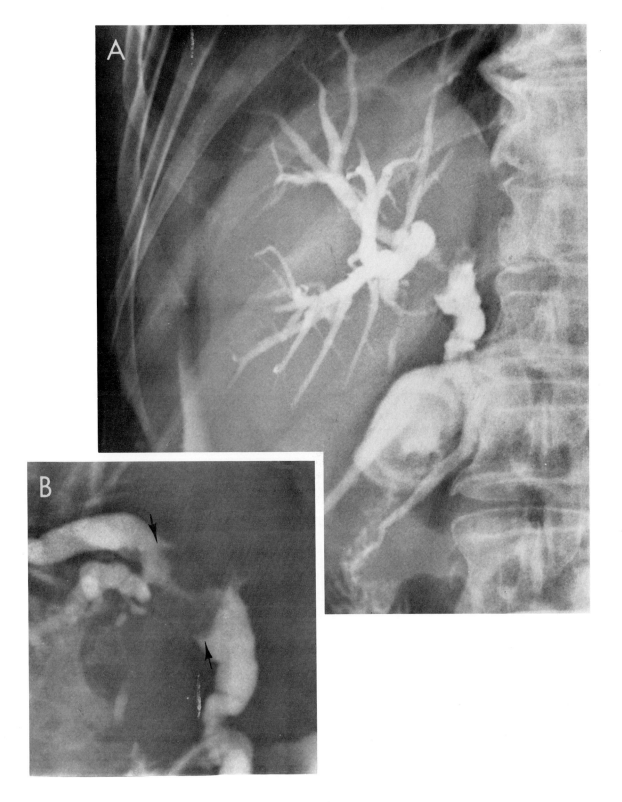

Figure 93 · Carcinoma of Common Hepatic Duct / 245

Figure 94.—Carcinoma of the common hepatic duct with extension to the right and left hepatic ducts.

A 51-year-old man with right upper quadrant pain had had a previous cholecystectomy and common bile duct exploration. The liver was palpably enlarged.

Percutaneous transhepatic cholangiogram, anteroposterior projection: There is an annular stenosis of the common hepatic duct with narrowing also of the right and left hepatic ducts at their junction. The lesion is annular and shows sharp overhanging margins at its distal end (**arrows**). Numerous opaque calculi are seen within the dilated common bile duct.

Comment: Cholelithiasis is found in 20% of cases of carcinoma of the bile ducts.

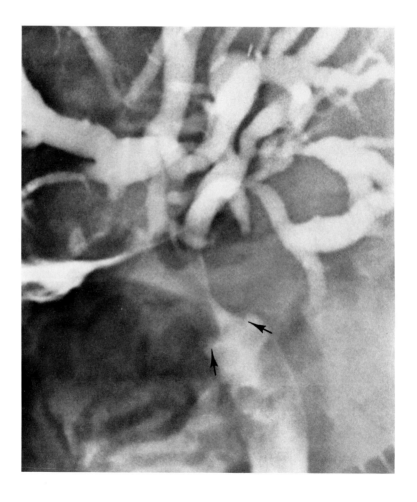

Figure 95.—Carcinoma of the common hepatic duct involving the right and left hepatic ducts.

A 57-year-old man had a 10-week history of gradual onset of jaundice and weakness. He had lost weight. The liver was enlarged 4 cm below the right costal margin and was nodular.

Percutaneous transhepatic cholangiogram, anteroposterior projection: Note a smooth, tapered narrowing of the distal right and left hepatic ducts near their confluence (**arrows**). Opaque calculi are present within the left hepatic duct. A minimal amount of contrast material passes distally into the common hepatic duct, but this is partially obscured by extravasated contrast material.

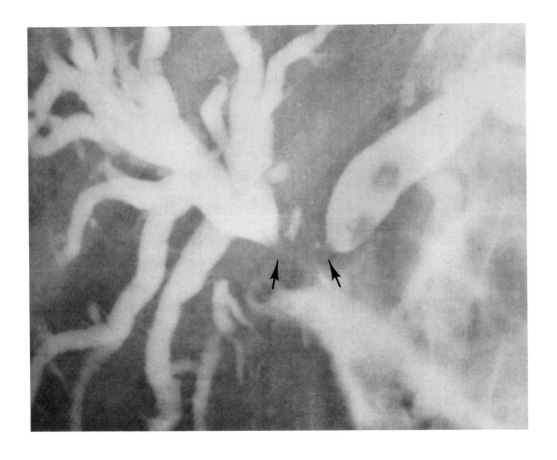

Figure 96.—Carcinoma of the common hepatic duct involving the right and left hepatic ducts with metastasis to the liver.

A 49-year-old woman had an onset of jaundice, right upper quadrant abdominal pain and anorexia two weeks before hospitalization. The liver was enlarged 3 cm below the right costal margin.

A, selective celiac angiogram, arterial phase: Marked narrowing of the right hepatic artery is present because of tumor encasement (**arrows**). The lateral border of the gastroduodenal artery (**arrowhead**) is irregular. No tumor vessel or tumor blush is demonstrated.

B, selective celiac angiogram, capillary phase: The early hepatogram shows a mottled parenchyma that is largely avascular. This indicates extensive metastasis within the liver.

Comment: The angiographic changes occurring with carcinoma of the bile ducts may be minimal. Blood vessel encasement, narrowing and obstruction occur most frequently. Extensive tumor vasculature and tumor blush are rare findings. Approximately one-third of patients have metastases at the time of examination.

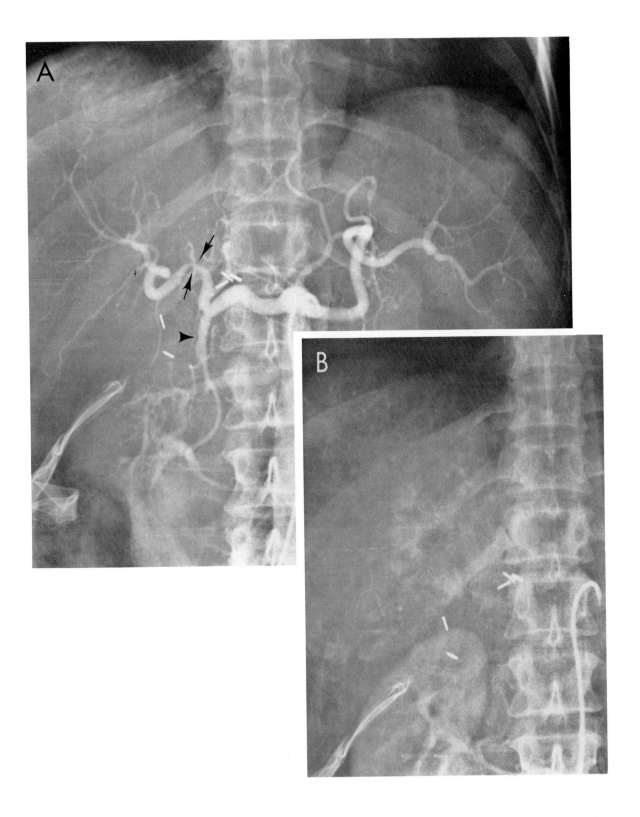

Figure 96 · Carcinoma of Hepatic Ducts / 249

Figure 97.—Carcinoma of the common hepatic duct and right and left hepatic ducts.

A 48-year-old woman had chronic abdominal pain and a 20-lb weight loss over a two-year period. Cholecystectomy with common bile duct exploration was performed one year before admission. Jaundice appeared two months prior to this study.

T-tube cholangiogram, anteroposterior projection: The hepatic duct is markedly stenosed, with smooth tapering to the point of obstruction (**arrows**). No contrast material passes proximally to the right hepatic duct. A minimal amount passed beyond the point of stenosis into the dilated left hepatic duct.

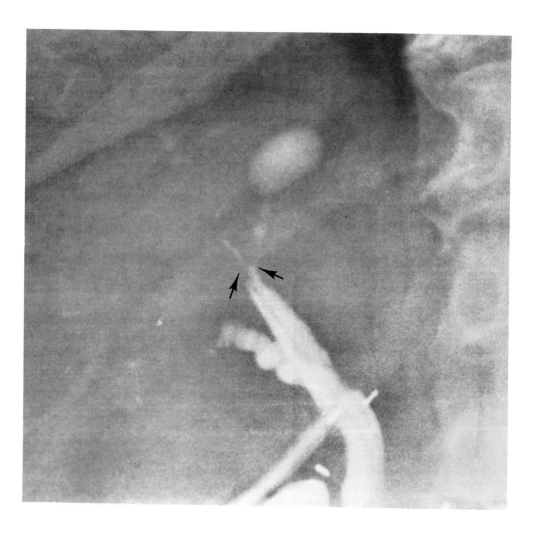

Figure 98.—Carcinoma of the common bile duct and left hepatic duct.

T-tube cholangiogram, anteroposterior projection: Irregularity of the medial wall of the common bile duct is seen opposite the point of insertion of the T-tube (**arrow**). The left hepatic duct is stenosed proximal to the most superior portion of the upper limb of the T-tube (**arrowhead**). The left hepatic duct proximal to the obstruction is dilated.

Comment: In this case, carcinoma is present in two separate foci. It is possible that one site represents extension of the tumor from the primary site.

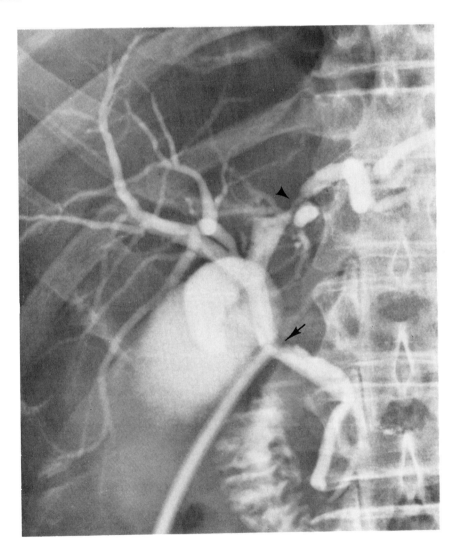

Figure 99.—Carcinoma of the common hepatic and right and left hepatic ducts.

A 73-year-old woman complained of nausea and vomiting and jaundice for two weeks. She was deeply jaundiced and cachectic.

T-tube cholangiogram, spot film, left anterior oblique projection: The right and left hepatic ducts are conspicuously narrowed at their confluence with the common hepatic duct; the common hepatic duct also is markedly narrowed at this point (**arrowheads**). There is minimal dilatation of the intrahepatic ducts proximal to the stenosis. The borders of the lesion are smoothly tapering.

Comment: The appearance of the biliary duct carcinoma in this case is typical of the type that grows largely intramurally. These tumors are usually small in volume and have a dense fibrous structure.

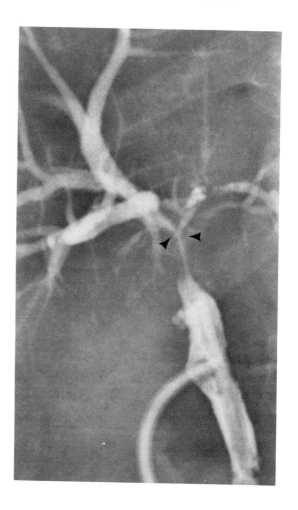

Figure 100.—Carcinoma of the right and left hepatic ducts and the proximal common hepatic duct.

A 60-year-old woman had a gradual onset of jaundice and abdominal discomfort over a two-month period.

T-tube cholangiogram, anteroposterior projection: The tip of the proximal limb of the T-tube is approximately at the junction of the right hepatic duct with the common hepatic duct. There is marked narrowing of the right hepatic duct and the proximal common hepatic duct in this region. The left hepatic duct is unfilled as a result of complete obstruction by the tumor. The intrahepatic branches of the right hepatic duct are moderately dilated. The inferior border of the right hepatic duct is irregular (**arrowheads**) and gradually tapers to its point of maximal stenosis.

Comment: The intramural scirrhous type of biliary duct carcinoma has a propensity for extending along the wall of the bile ducts. In this case the tumor involves quite a long segment of the wall of the right hepatic duct, as indicated by the irregularity of the inferior margin.

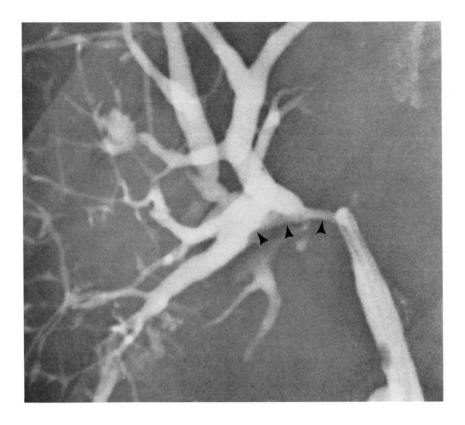

Figure 101.—Carcinoma of the ampulla of Vater.

A 53-year-old woman had gradual progression of jaundice over a three-week period.

Hypotonic duodenogram, spot film, anteroposterior projection: A large, smooth and well-defined polypoid mass extends intraluminally from the medial wall of the duodenum in the region of the ampulla (**arrowheads**). Size of lumen and mucosal pattern of the duodenum appear to be normal.

Comment: An ampulla of Vater measuring more than 1.5 cm in diameter should be viewed with suspicion. Other pathologic change may not be demonstrated. Apart from the increase in size, the papilla may have a normal, regular and smooth appearance. When the tumor is extensive, involvement of the duodenum will be demonstrated. Involvement of the duodenal loop may be manifested on the routine gastrointestinal study or on hypotonic duodenography as extrinsic pressure on the medial wall of the duodenum causing fold effacement, indentation of the medial wall as a result of the mass, and an inverted-figure-3 sign.

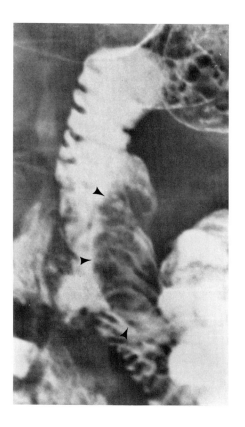

Figure 102.—Carcinoma of the ampulla of Vater.

This 79-year-old man had noted anorexia, weakness and a 20-lb weight loss over a period of four months. Shortly before hospitalization, he had become jaundiced.

Upper gastrointestinal radiograph, prone oblique spot film: An intraluminal mass lesion (**arrowheads**) is seen projecting from the medial aspect of the duodenal loop. The mass is smoothly lobulated.

Comment: Radiographs in numerous projections are required for examination of the duodenum, particularly when an ampullary lesion is suspected. In this case, the tumor was demonstrable only when the duodenal loop was fully distended with barium. Hypotonic duodenography is indicated when lower bile duct obstruction is suspected. This ampullary enlargement may also be noted in cases of impacted biliary calculi in the lower common bile duct. For this reason, interpretation of the enlarged ampulla in obstructive jaundice can prove difficult.

Figure 102 · **Carcinoma of Ampulla of Vater**

Figure 103.—Carcinoma of the ampulla of Vater.

A 40-year-old man had three episodes of epigastric pain radiating to the back and vomiting and jaundice over a 17-month period. Surgical or pathologic proof of the diagnosis was not available.

A, intravenous cholangiogram, right posterior oblique projection: There is only fair opacification of the common bile duct. The common bile duct and the common hepatic duct are enlarged (**arrowheads**). The contrast material was retained in the common bile duct for an extended period, indicating partial obstruction.

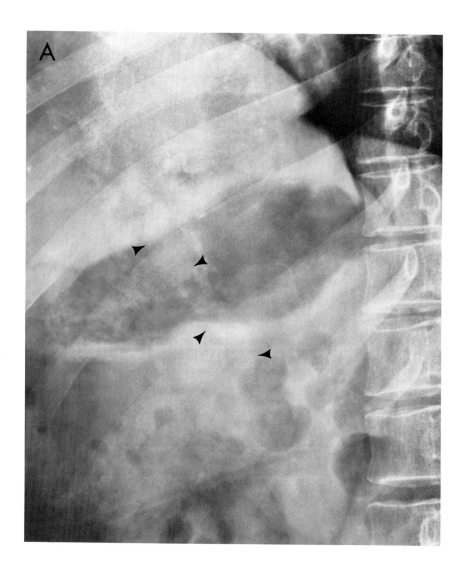

B, upper gastrointestinal radiograph, multiple spot films of the duodenal loop: A smooth, well-defined mass protrudes intraluminally from the medial wall of the duodenum in the region of the ampulla of Vater (**arrowheads**). The appearance is quite characteristic of ampullary carcinoma.

Comment: Although this case has not been proved, clinical and radiologic findings are characteristic of carcinoma of the ampulla. Partial obstruction of the lower common bile duct plus a smooth, well-defined mass in the region of the ampulla are strong radiographic evidence of ampullary carcinoma.

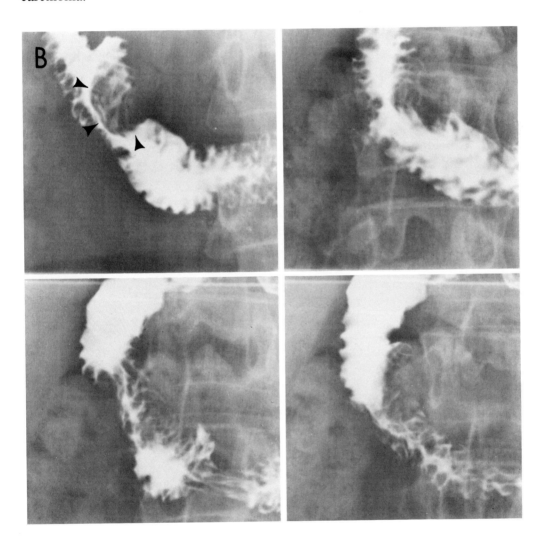

Figure 104.—Carcinoma of the ampulla of Vater.

A 68-year-old man had pruritus of 1 month's duration accompanied by gradual onset of weakness, anorexia and malaise. He had a recent onset of jaundice, and the liver was enlarged. Cirrhosis had been diagnosed 10 years previously.

A, upper gastrointestinal radiograph, left posterior oblique spot film of the proximal duodenum: There is evidence of extrinsic pressure on the duodenum in its immediate postbulbar portion. A small, nodular luminal filling defect is apparent distally in the duodenum in the region of the ampulla of Vater (**arrows**).

B, percutaneous transhepatic cholangiogram, right posterior oblique projection: The common bile duct, the common hepatic duct and the intrahepatic ducts are grossly dilated. A minimal quantity of barium outlines the duodenal loop. The extrinsic impression on the postbulbar duodenum, as seen in **A,** corresponds to the position of the dilated common bile duct. The roughly circular filling defect is incompletely surrounded by contrast material in the distal common bile duct at the level of the ampulla.

(*Continued.*)

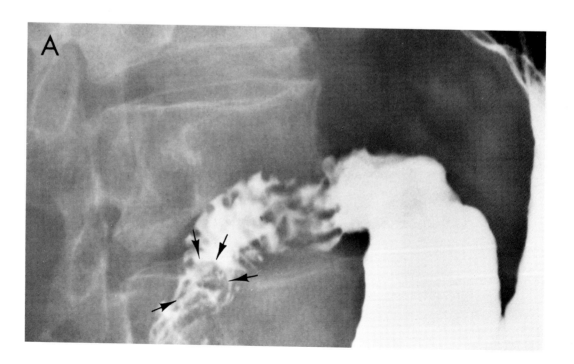

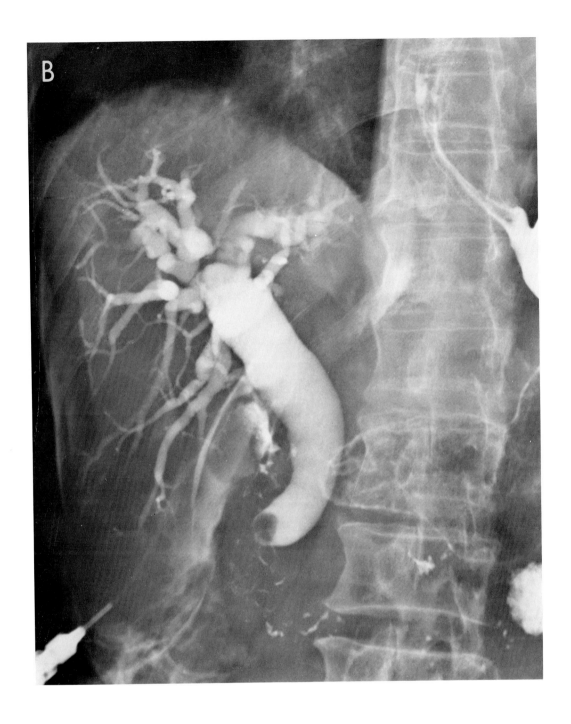

Figure 104 · Carcinoma of Ampulla of Vater / 259

Figure 104 (cont.).—Carcinoma of the ampulla of Vater.

C, percutaneous transhepatic cholangiogram, right posterior oblique spot film: The filling defect in the distal common bile duct is now clearly defined. The lesion completely obstructs the common bile duct. Contrast material does not surround the obstructing lesion in its entirety. The margins of the filling defect are slightly irregular.

Comment: The findings on percutaneous cholangiography in this case are unusual. The largely intraluminal character of the obstructing lesion in the common bile duct with a convex meniscus suggests an impacted calculus. Enlargement of the ampulla of Vater can be found in impacted calculi attributed to edema. The greatly enlarged common bile duct is usually not the result of calculi, as the ducts often are not distensible because of previous inflammation. Such enlargement of the bile ducts is more frequently found in carcinoma.

Figure 104, courtesy of Dr. M. S. Kleinman, Kaiser Foundation Hospital, Los Angeles.

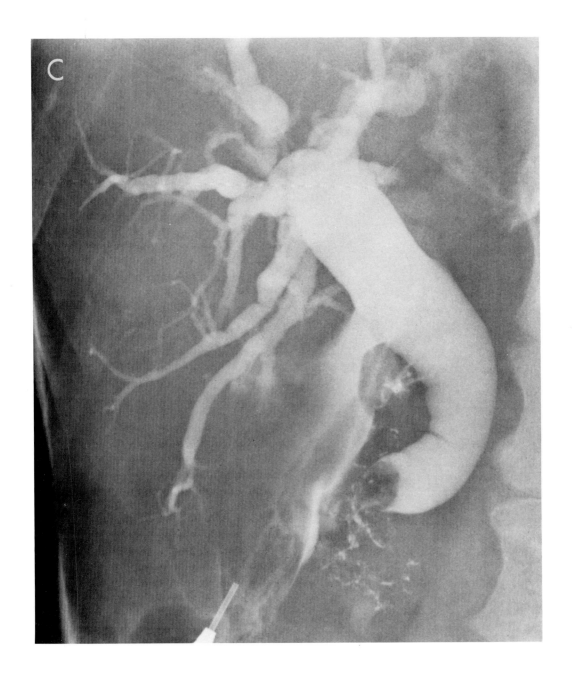

Figure 104 · Carcinoma of Ampulla of Vater / 261

Figure 105.—Villous adenoma of the ampulla of Vater.

A, upper gastrointestinal radiograph, anteroposterior spot film of duodenal loop: A smooth, marginal filling defect is present in the second portion of the duodenum in its medial aspect (**arrowheads**). This indicates enlargement of the ampulla of Vater.

B, intravenous cholangiogram, right posterior oblique projection at 120 minutes: The common bile duct is markedly narrowed immediately proximal to the papilla (**arrowheads**). The narrowed segment is enclosed in the enlarged ampulla, which is outlined by contrast material within the duodenum and common bile duct.

C, retrograde pancreatic ductogram, lateral projection: Contrast material has been injected into the duct of Wirsung after cannulation of the papilla. Contrast material within the duodenal loop outlines the enlarged ampulla of Vater (**arrows**). The segment of the duct within the enlarged ampulla (**arrowheads**) is conspicuously narrowed. The common bile duct was not filled.

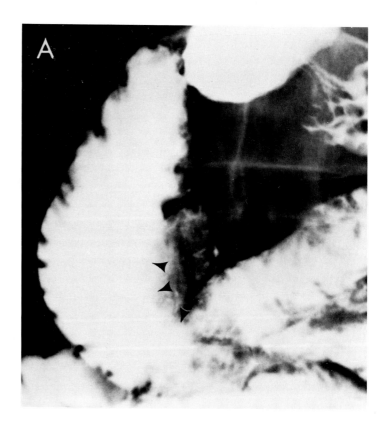

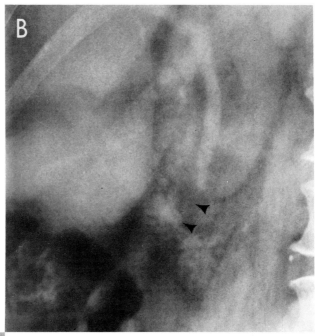

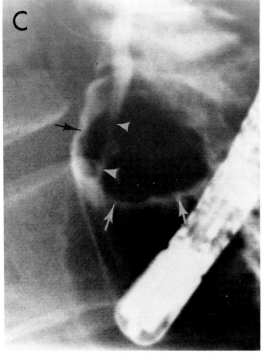

Figure 105 · Adenoma of Ampulla of Vater / 263

Figure 106.—Choledochocele.

A man had had a cholecystectomy and common bile duct exploration. On this occasion he had right upper quadrant abdominal pain and jaundice.

A, upper gastrointestinal radiograph, anteroposterior projection: A smooth intraluminal filling defect is present in the second portion of the duodenum arising from the medial wall (**arrowheads**). This has the appearance of a smooth enlargement of the ampulla of Vater.

B, T-tube cholangiogram, left anterior oblique projection: The common hepatic duct and the common bile duct are moderately dilated. There is a saccular dilatation of the distal common bile duct, and this segment of the common bile duct has prolapsed into or invaginated the medial wall of the duodenum (**arrowheads**). This prolapsed segment corresponds to the filling defect noted in **A.** The common bile duct is partially obstructed.

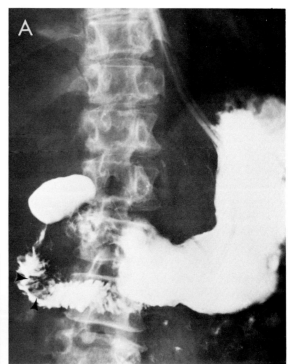
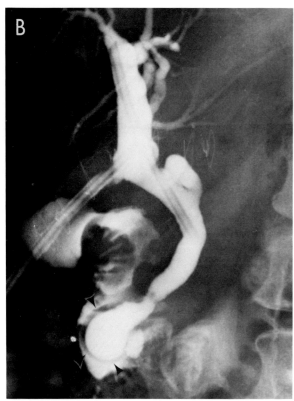

Figure 106 · Choledochocele / 265

Figure 107.—Choledochal cyst.

A young woman complained of intermittent upper abdominal pain. She had previously had a cholecystectomy. A choledochal cyst containing numerous calculi was found at operation.

A, intravenous cholangiogram, right posterior oblique projection at 120 minutes: The common hepatic duct and the proximal bile duct are greatly dilated. A large circular opacity with multiple lucencies within is outlined (**arrowheads**). The distal common bile duct is not delineated.

B, intravenous cholangiogram, laminagram: The circular, opaque-filled structure lying medial to the partially obstructed common bile duct is more clearly defined. This cystic structure has many lucent areas indicating biliary calculi within it.

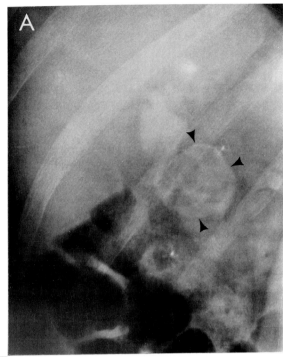
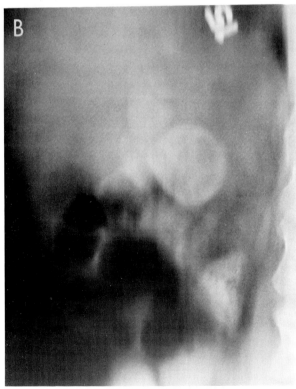

Figure 107 · Choledochal Cyst / 267

Figure 108.—Compression and narrowing of the common hepatic duct due to metastasis from carcinoma of the stomach secondary to nodes in the hilus of the liver.

This 52-year-old woman had a gastric resection 3 years previously for carcinoma. She had noted pruritus for a 3-month period. Jaundice was first noticed 10 days before hospitalization.

Percutaneous transhepatic cholangiogram, anteroposterior spot film: The common hepatic duct is compressed by an extrinsic mass lying superiorly and medially (**arrowheads**). The intrahepatic ducts are markedly distended.

Comment: Metastasis to nodes in the hilus of the liver may be large enough to cause biliary obstruction, and their appearance on cholangiograms must be differentiated from primary neoplasms of the bile ducts. The length and smooth character of the lesion and lack of intrinsic involvement of the bile ducts are characteristic.

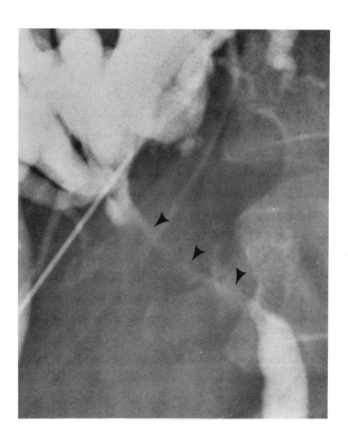

Figure 109.—Chronic inflammatory granulation tissue in the common bile duct having the appearance of a nodular tumor.

A 62-year-old woman had two biliary fistulas develop after cholecystectomy. One of the fistulas discharged bile; a second operation revealed chronic inflammatory changes in the common hepatic and the common bile ducts.

Sinus tract injection, anteroposterior projection: Contrast material was injected by way of a catheter inserted in one of the fistulous tracts that developed after cholecystectomy. This communicates with the biliary tree, and there is good opacification of the bile ducts. Multinodular filling defects simulating calculi of the common hepatic duct and proximal common bile duct are demonstrated (**arrowheads**).

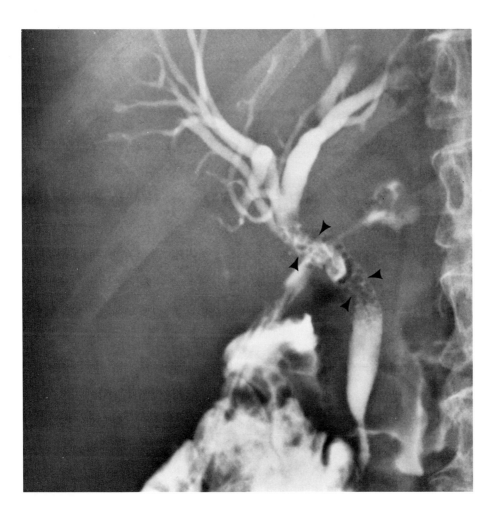

Figure 110.—Normal and dilated common bile ducts demonstrated by endoscopic cannulation and injection of contrast material.

A, peroral retrograde pancreatogram, right posterior oblique projection: The cannula tip has been inserted through the papilla and aqueous contrast material has been injected to opacify the biliary ducts. The duct of Wirsung is also opacified. The common bile duct is normal in appearance.

(Continued.)

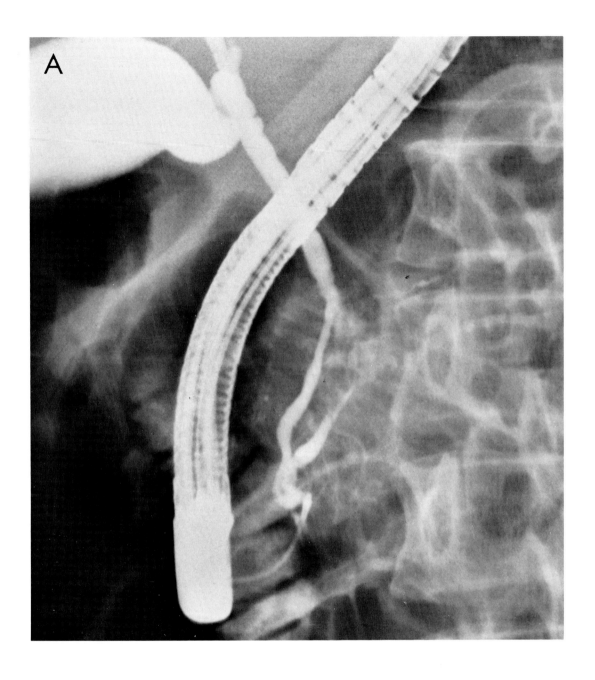

Figure 110 · Normal & Dilated Common Bile Ducts

Figure 110 (cont.).—Normal and dilated common bile ducts demonstrated by endoscopic cannulation and injection of contrast material.

B, peroral retrograde pancreatogram, anteroposterior projection: The tip of the cannula is within the common bile duct and contrast material has been injected. There is gross dilatation of the common bile duct, but it otherwise appears normal. The obstructing lesion is not identified, and the duct of Wirsung is not opacified.

Comment: Increased diagnostic capability is lent by the addition of the fiberoptic examination in the assessment of biliary and pancreatic pathology. This examination is especially useful when laboratory studies indicate probable nonopacification with intravenous and oral cholangiographic contrast agents and when transhepatic cholangiography is unsuccessful or contraindicated. The discomfort and risk to the patient with retrograde ductography in experienced hands is less than that with percutaneous transhepatic cholangiography. One study reports a success rate of 86% in cannulization of the papilla, with the pancreatic duct demonstrated in 83% of attempts. The pancreatic duct was never delineated consistently preoperatively before the advent of fiberoptic cannulation and retrograde ductography.

Figure 110, courtesy of Dr. Itaru Oi, Institute of Gastroenterology, Tokyo Women's Medical College, Tokyo, Japan.

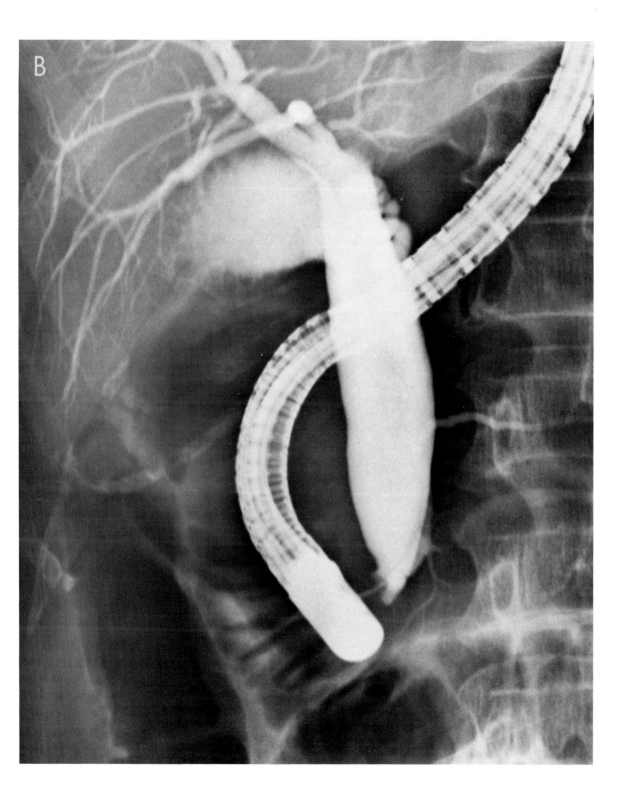

Figure 110 · Normal & Dilated Common Bile Ducts

PART 4

The Gallbladder

Characteristics of Tumors of the Gallbladder

CARCINOMA

Carcinoma of the gallbladder accounts for approximately 3% of the deaths related to cancer. Cholelithiasis bears an important relationship to primary carcinoma of the gallbladder and is present in about 90% of the patients. Although derivatives of cholic acid are powerful carcinogens, carcinoma of the gallbladder is an uncommon form of cancer. It is possible that chronic irritation and infection, which frequently occur in conjunction with cholelithiasis, are causative factors as important as cholic acid derivatives. The incidence of carcinoma of the gallbladder in macroscopic gallbladder specimens varies from 0.5–3%.

The occurrence of carcinoma of the gallbladder, like cholelithiasis, is more frequent in women than in men, the ratio being 4:1. It appears most frequently in the sixth and seventh decades, and the average age at onset is 55 years.

PATHOLOGY.—Approximately 95% of the cancers of the gallbladder are adenocarcinomas, and 5% are squamous cell carcinomas. The most common growth pattern is infiltrative. Some are scirrhous in type and a small percentage are papillary. Often the site of origin of the tumor is not easily discernible either at operation or on pathologic examination. When the tumor is found at an early stage, the fundus and neck are the most commonly involved areas. In most instances, metastasis has occurred at the time the initial diagnosis of carcinoma of the gallbladder is made. The most frequent types of metastases are direct extension and lymphatic spread. Anterograde lymphatic spread occurs to the cystic and pericholedochal nodes; retrograde lymphatic spread occurs to nodes at the hilus of the liver. Metastasis to regional nodes occurs in approximately 75% of the patients. Direct extension to the liver is found in 50–62% of carcinomas of the gallbladder. Other neighboring organs may be involved, including the stomach, duodenum and colon. Disseminated liver metastases are rare but can occur after the tumor invades the portal vein.

CLINICAL FEATURES.—Most carcinomas of the gallbladder are diagnosed when operation is performed for cholelithiasis or cholecystitis. The correct preoperative diagnosis is made only in 5–6% of patients. Biliary colic and flatulent dyspepsia are frequent presenting symptoms, but these may mimic chronic inflammatory disease of the gallbladder. A recent change in character of the symptoms is an important indicator. In particular, recent weight loss plus sudden onset of constant pain should arouse suspicion. On physical examination the patient may present a palpable mass, an enlarged liver or ascites. Jaundice occurs in approximately 50% of patients.

Radiologic features.—Findings on conventional radiographs are infrequent and nonspecific. An enlarged liver or a right upper quadrant abdominal soft tissue density contiguous with the liver may be observed. Opaque calculi may be visible, but these give no clue to the coincidental malignancy. The frequency of carcinoma is greater in porcelain gallbladders. Calcification in the wall of the gallbladder should alert the radiologist to the possibility of an accompanying carcinoma.

Elevation of the right hemidiaphragm with associated right basilar atelectasis or small pleural effusion may occur. Barium studies may show deformity of the stomach, duodenum or colon as a result of pressure or extension into these organs.

The cystic artery is delineated in the great majority of celiac angiograms. Its most common origin is from the inferior aspect of the common hepatic artery immediately distal to the takeoff of the gastroduodenal artery. The medial and lateral branches of the cystic artery surround the gallbladder, and their separation on the angiogram indicates enlargement of the gallbladder. A slight blush during the capillary phase of the angiogram outlines the gallbladder wall. The venous drainage of the gallbladder may communicate with the hepatic veins, but usual communication is with the portal vein. The most frequent angiographic findings in carcinoma of the gallbladder are those associated with its propensity for infiltration of surrounding structures; arterial displacement and encasement are the major findings. The arteries most commonly involved are the cystic artery and its branches, the common and right hepatic arteries and medial branches of the right hepatic artery, and the gastroduodenal artery and its branches. Occasionally, tumor neovasculature will be apparent in the region of the gallbladder bed. The vessels are small, tortuous and irregularly distributed. Complete obstruction of vessels occurs; in particular, the cystic artery or its branches may be demonstrated incompletely. Tumor stain is uncommon and tumor vascularity, when present, is sparse. Invasion, thrombosis and obstruction of the portal vein have been reported. Early filling of hepatic veins has been noted. In general, carcinoma of the gallbladder is rarely diagnosed angiographically prior to dissemination of the tumor to an inoperable degree.

Sarcomas

Sarcomas are extremely rare tumors of the gallbladder. Spindle cell sarcoma, leiomyosarcoma and lymphosarcoma, primarily occurring in the gallbladder, have been reported. Their clinical and radiologic characteristics differ little, if any, from those previously described for carcinoma of the gallbladder.

Metastatic Tumors

Metastatic tumors of the gallbladder are extremely rare. The most common metastatic tumor of the gallbladder is malignant melanoma. Metastases in the gallbladder may be in the form of implants in the serosa or they may be mucosal in location. Some may be polypoid. The liver is involved in the great majority of instances. Cases of metastatic melanoma appearing on the cholecystogram as multiple fixed filling defects within the gallbladder have been reported.

Polyps

The radiologic diagnosis of polyps of the gallbladder comprises a variety of pathologic lesions, including papilloma, adenoma, cholesterol polyp, adenomyoma and inflammatory polyp.

Papilloma.—Papilloma occurs approximately once in every 100 cholecystectomies. The papilloma is a true neoplastic lesion consisting of a focal epithelial proliferation; the columnar epithelium and vascular connective tissue within have a villous structure. They may be single or multiple, sessile or pedunculated and are frequently less than 0.5 cm in diameter. Cholelithiasis is present in 68% of the patients.

Adenoma.—This polyp is most commonly found in the fundus of the gallbladder. It is a true adenoma, having a glandular structure like that found in adenomas elsewhere in the gastrointestinal tract. Secretory glands occur most frequently in the fundus of the normal gallbladder; this accounts for the occurrence of the adenoma in this location. This polyp may also be sessile or pedunculated. It usually has a smooth, velvety surface and most frequently occurs singly. Adenomas often measure between 0.5 and 4 cm in diameter. Gallstones are coincident in 38%.

Cholesterol polyps.—These arise from foci of cholesterosis. Cholesterol is deposited in lipid-laden macrophages beneath the basement membrane of the columnar epithelium. These polyps may be sessile or pedunculated and are frequently on a long slender stalk that may be detached easily. They are generally multiple. It has been proposed that when these polyps become detached from the gallbladder wall, they form the nidus around which a calculus develops.

Adenomyoma.—This is a tumor of developmental origin. It is frequently considered to be a hamartoma. It usually occurs in the fundus, is sessile and is usually single. On macroscopic examination the polyp exhibits central umbilication. Microscopically, it is composed of tubular glandular structures within the mucosal and submucosal layers of the gallbladder with interlacing smooth muscle bundles between the tubular structures.

Inflammatory polyps.—An inflammatory polyp is a localized projection of proliferating mucosa with edema and chronic inflammatory cells plus glandular structures caused by a chronic inflammation of the gallbladder termed cholecystitis glandularis proliferans. The chronic inflammation is also reflected in the thickening of the gallbladder wall. Polyps are frequently associated with gallbladder calculi.

Some other extremely rare lesions presenting as tumors of the gallbladder have been reported, including mucous or epithelial cysts of the gallbladder wall, carcinoid, pancreatic rest and neurinoma. Others are myomas, lipomas, myxomas and fibromas.

CLINICAL FEATURES OF POLYPS.—Gallbladder polyps are usually asymptomatic. When symptoms such as biliary colic, nausea and vomiting, intolerance to fatty foods and flatulence occur, they are most frequently ascribable to coincidental cholelithiasis or chronic cholecystitis. Occasionally, biliary symptoms have been attributed to gallbladder polyps in the absence of other disease of the gallbladder, but this is a dangerous concept.

RADIOLOGIC FEATURES.—A filling defect in the opacified gallbladder which maintains its position within the gallbladder on different projections or on repeated studies is characteristic of a polyp of the gallbladder. In general, polyps of the gallbladder cannot be differentiated radiologically with the possible exception of the adenomyoma, which may appear as a fundal lesion with a central depression, particularly on the radiograph of the contracted gallbladder following the ingestion of a fatty meal. The majority of lesions are 2–3 mm in diameter. The cholecystographic medium is usually well concentrated within the gallbladder. Most polyps are marginal and as well as maintaining their position within the gallbladder maintain the same position relative to one another when multiple. They are best seen in the contracted gallbladder. A notch in the outline of the gallbladder at the base of the polyp stalk is an occasionally seen confirmatory sign.

SUGGESTED READING

Berk, R. N.; Armbuster, T. G., and Saltzstein, S. L.: Carcinoma in the porcelain gallbladder, Radiology 106:29–31, January, 1973.

Cohn, E. M.: Tumors of the Gallbladder and Bile Ducts, *in* Bockus, H. L. (ed.): *Gastroenterology* (Philadelphia: W. B. Saunders Company, 1965), Vol. III, pp. 811–820.

Jutras, J. A., and Levesque, H. P.: Adenomyoma and adenomyomatosis of the gallbladder: Radiologic and pathologic correlations. Radiol. Clin. North America 4:483–500, December, 1966.

Nugent, F. W.; Meissner, W. A., and Hoelscher, F. E.: The significance of gallbladder polyps, J.A.M.A. 178:426–428, October 28, 1961.

Ochsner, S. F.: Solitary polypoid lesions of the gallbladder, Radiol. Clin. North America 4:501–510, December, 1966.

Ochsner, S. F.: Intramural lesions of the gallbladder, Am. J. Roentgenol. 113:1–9, September, 1971.

Rösch, J.; Grollman, J. H., Jr., and Steckel, R. J.: Arteriography in the diagnosis of gallbladder disease, Radiology 92:1485–1491, June, 1969.

Shimkin, P. M.; Soloway, M. S., and Jaffe, E.: Metastatic melanoma of the gallbladder, Am. J. Roentgenol. 116:393–395, October, 1972.

Sprayregen, S., and Messinger, N. H.: Carcinoma of the gallbladder: Diagnosis and evaluation of regional spread by angiography, Am. J. Roentgenol. 116:382–392, October, 1972.

Figure 111.—Carcinoma of the gallbladder.

A, selective celiac angiogram, arterial phase: Note narrowing and encasement of the right hepatic artery immediately proximal to its primary branches (**arrowheads**). The cystic artery, which has its origin from the right hepatic artery in this instance, is tortuous and its branches do not exhibit their usual smooth, curved course. Minimal irregularity is seen on the main trunk of the cystic artery (**arrow**). The inferior branches of the right hepatic artery are displaced superolaterally.

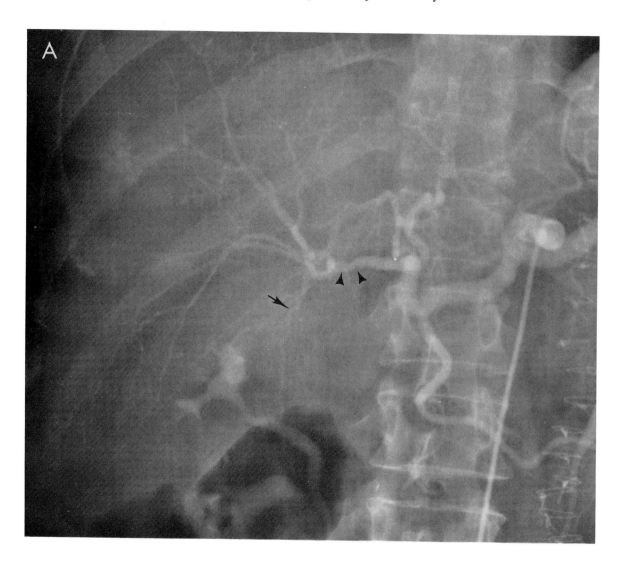

B, selective celiac angiogram, portal venous phase: The right branch of the portal vein is displaced superiorly and there is minimal irregularity of the inferior border of the vein where it is displaced (**arrowheads**). In the region of the tumor, which has arisen in the gallbladder and has infiltrated the liver, the portal hepatogram is deficient.

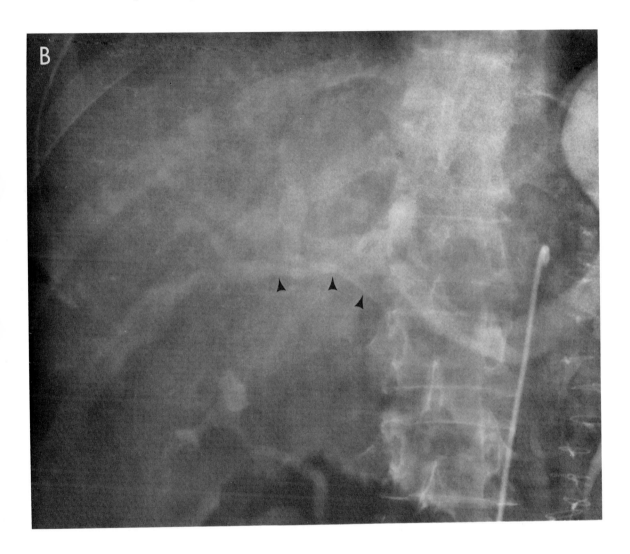

Figure 112.—Carcinoma of the gallbladder.

A 70-year-old man complained of anorexia, weakness, weight loss and right upper quadrant abdominal pain of two months' duration. The liver was enlarged 3 cm below the right costal margin.

Selective celiac angiogram, arterial phase: Minimal irregularity of the lower border of the right hepatic artery is seen. The branches of the cystic artery (**arrowheads**) appear to be minimally separated, with a faint blush of contrast medium in the area encompassed by these arteries. The main trunk of the cystic artery is obscured. The gastroduodenal artery and its branches are incompletely filled on this injection, but they were opacified on the superior mesenteric angiogram. The proper hepatic artery and the right hepatic artery are of large caliber.

Comment: The irregularity of the right hepatic artery and the minimal tumor blush encompassed by the branches of the cystic artery are evidence of carcinoma of the gallbladder with local extension.

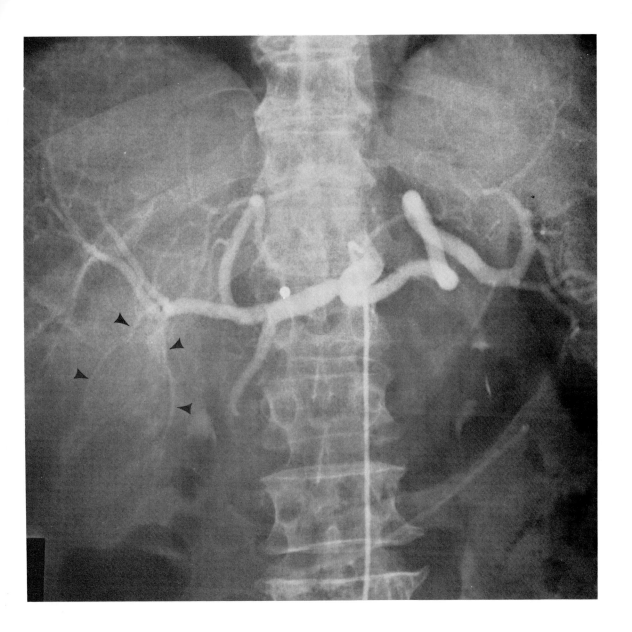

Figure 113.—Carcinoma of the gallbladder.

A 68-year-old woman had anorexia and lost 14 lb over a period of six months. The urine had become increasingly darker over a period of five weeks and jaundice had been noted for one week.

A, spot film of the duodenal loop, right anterior oblique projection: A smooth extrinsic pressure defect is demonstrated on the superior aspect of the first part of the duodenum (**arrows**).

B, superior mesenteric angiogram, late arterial phase: The proper hepatic artery arises from the superior mesenteric artery, with marked narrowing of the right hepatic artery at its origin from the proper hepatic artery (**arrow**). Good filling of the branches of the right hepatic artery is demon-

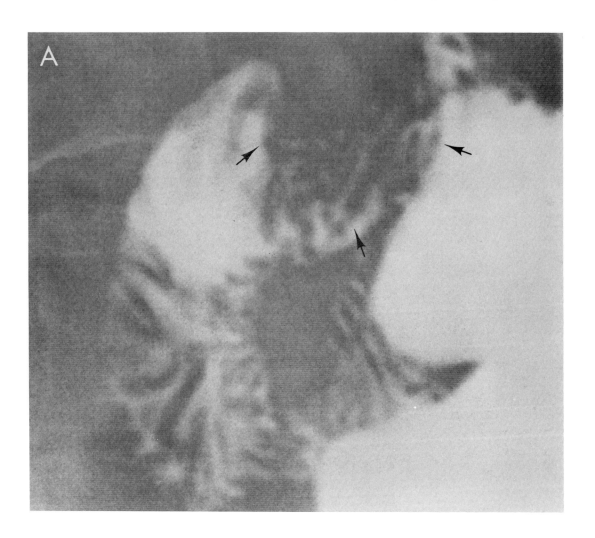

strated; the inferior branches are displaced laterally and draped around an avascular mass pressing on the inferomedial margin of the liver (**arrowheads**). The cystic artery is not demonstrated.

Comment: The carcinoma of the gallbladder in this patient had spread to the cystic duct and common bile duct and had caused biliary obstruction. The common hepatic duct was widely dilated. These findings account for the extrinsic pressure defect on the superior aspect of the first portion of the duodenum. The tumor encasement of the proximal right hepatic artery is typical of carcinoma of the gallbladder, although cholangiocarcinoma could present similarly.

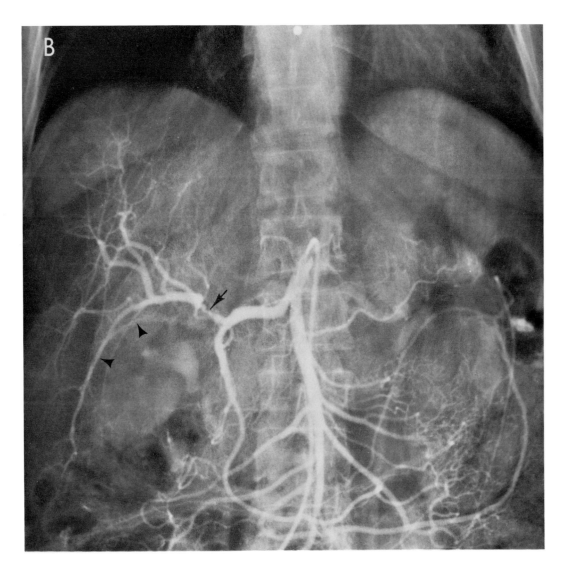

Figure 114.—Carcinoma of the gallbladder.

A 71-year-old woman had indigestion, heartburn and belching for one year with the recent onset of jaundice. The liver was palpably enlarged.

A, upper gastrointestinal radiograph, right anterior oblique projection: Note the extrinsic pressure defect on the lateral aspect of the second portion of the duodenum (**arrows**).

B, upper gastrointestinal radiograph, left posterior oblique projection: There is pressure on the superior aspect of the distal antrum and first por-

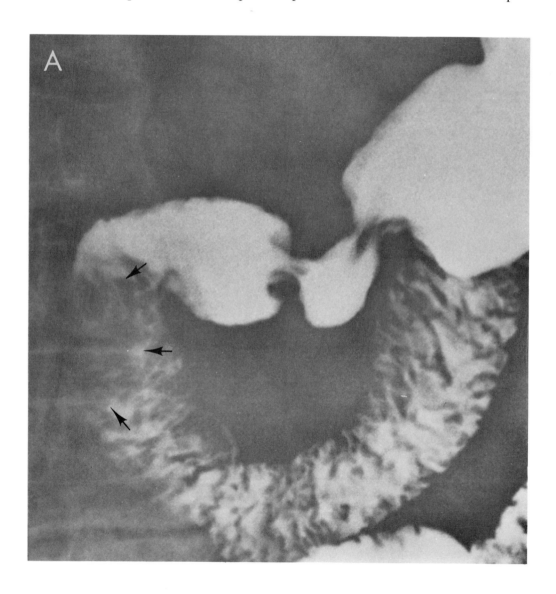

tion of the duodenum by the enlarged liver. The lateral aspect of the proximal descending portion of the duodenum appears to be fixed and rigid, with stretching and distortion of the mucosa and probable mucosal ulceration in that region (**arrows**).

Comment: Carcinoma of the gallbladder is widely disseminated at the time of diagnosis in most instances. In this patient the tumor had extended medially and invaded the wall of the duodenum. Calculi were present in the gallbladder.

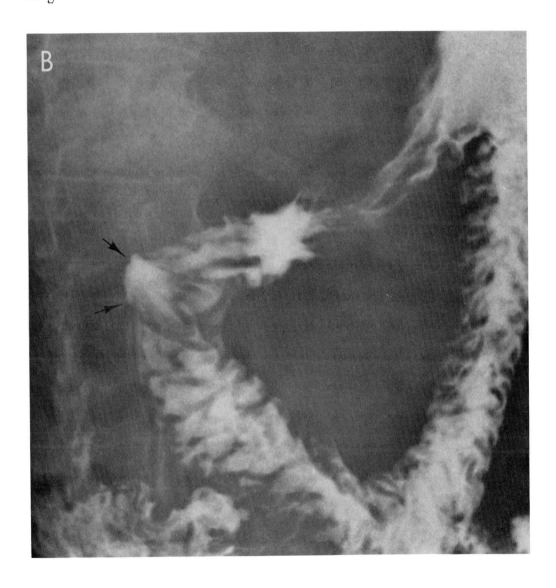

Figure 115.—Carcinoma of the gallbladder.

This patient was admitted for further evaluation after a laparotomy at another institution where the diagnosis of carcinoma of the gallbladder with extension to the liver was made.

A, selective celiac angiogram, arterial phase: There is wide splaying of the branches of the cystic artery (**arrowheads**). The area from the medial branch of the right hepatic artery to the lower portion of the right lobe (**arrow**) is markedly tortuous and displaced laterally, showing variation in lumen size and abnormal branching. Tumor neovasculature is not demon-

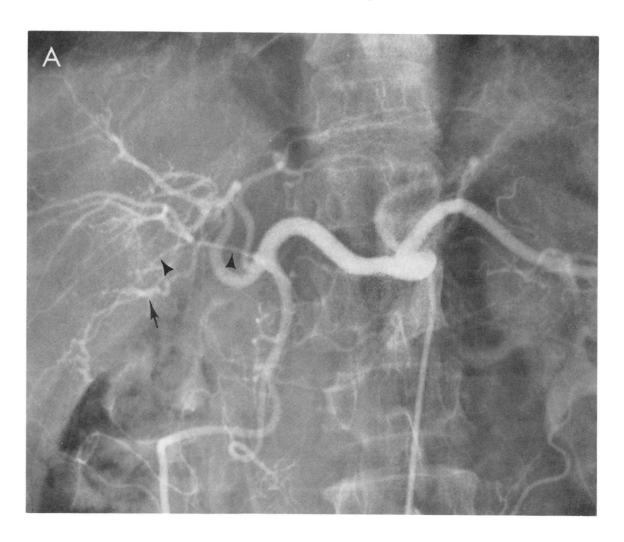

strated, and tumor blush is not shown. The gastroduodenal artery is displaced medially.

B, liver scan, anteroposterior supine projection: A large area of decreased activity is present in the region of the gallbladder bed. In this patient the defect is much larger than the expected anatomic defect as a result of the normal gallbladder.

Comment: In this patient the propensity of carcinoma of the gallbladder for invasion of surrounding organs is well demonstrated by the involvement of the right hepatic artery with tumor. The size of the mass can be appreciated by the marked splaying of the branches of the cystic artery.

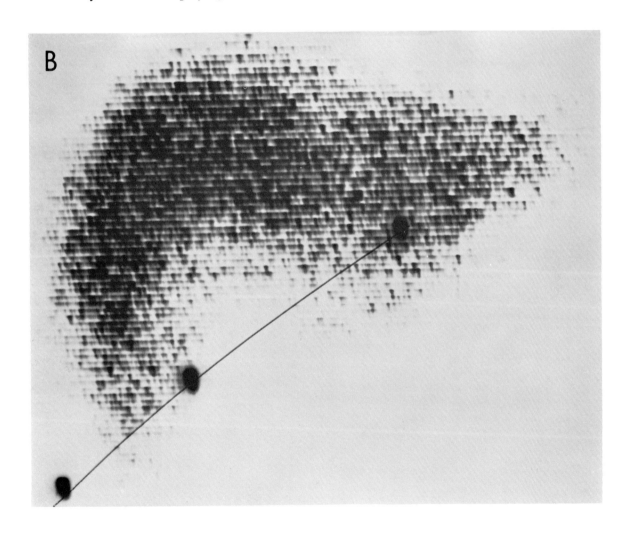

Figure 116.—Carcinoma of the gallbladder.

A, oral cholecystogram, posteroanterior oblique projection: The gallbladder is moderately well opacified, and a large filling defect is present in the midbody contiguous with the lateral wall of the gallbladder (**arrowheads**). Many smaller lucencies are suggested in the fundus of the gallbladder.

B, oral cholecystogram, posteroanterior projection: The gallbladder is now viewed in a different projection, but the lucency in association with the lateral wall maintains the same position relative to the wall (**arrowheads**) as noted in **A**.

C, oral cholecystogram, posteroanterior projection after a fatty meal: Multiple small lucencies are now defined within the fundus of the gallbladder. The larger lucency related to the lateral wall of the gallbladder remains constant.

Comment: Opacification of the gallbladder after ingestion of oral cholecystographic medium is the exception rather than the rule with carcinoma of the gallbladder. Polypoid carcinomas are the rarest of those affecting the gallbladder; the majority are infiltrating or scirrhous. Carcinoma in situ has been discovered in a number of instances after cholecystectomy for a polyp of the gallbladder. It is debatable whether all patients with an asymptomatic polyp should undergo cholecystectomy. It is estimated that the mortality rate from cholecystectomy about equals the chance of developing carcinoma in a polyp.

Figure 116, courtesy of Dr. Sidney Rubin, Kansas City, Mo., and Eastman Kodak Co., Rochester, N.Y.

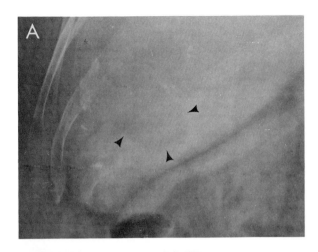
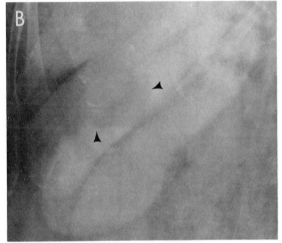
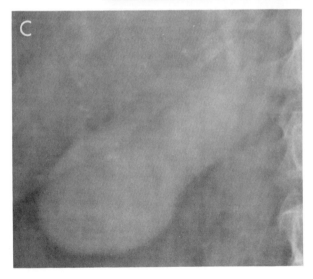

Figure 116 · Carcinoma / 293

Figure 117.—Carcinoma of the gallbladder.

Oral cholecystogram, posteroanterior projection: The gallbladder is well opacified, with a large marginal defect along the medial wall, projecting intraluminally. No gallstones are present. The defect maintained its position on several views.

Comment: The smoothness of the outline of this polypoid lesion suggests a benign rather than a malignant tumor. The tumor in this 64-year-old woman had extended through the wall of the viscus, with metastatic nodules in the liver. The dense opacification of the gallbladder and absence of cholelithiasis is remarkable.

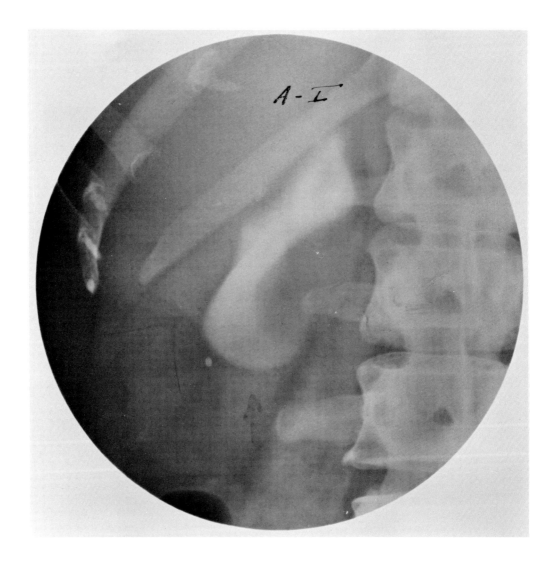

Figure 118.—Adenoma of the gallbladder.

A 66-year-old man had complained of epigastric postprandial pain for one year.

A, oral cholecystogram, spot film, left anterior oblique projection: Associated with the superolateral wall of the gallbladder is a moderately large filling defect (**arrowheads**) with indentation of the gallbladder wall at this point. A second filling defect is suggested at the fundus of the gallbladder.

B, oral cholecystogram, left anterior oblique projection after a fatty meal: The gallbladder is now contracted, and the opacification is more dense than in **A**. The filling defect at the superolateral margin of the gallbladder remains constant (**arrowheads**), while the filling defect at the fundus is not now visible and the suggested defect in **A** may have been the result of gas in the bowel.

Comment: The radiograph after a fatty meal is most useful for confirmation of polyps of the gallbladder. The contraction of the gallbladder appears to accentuate the filling defect caused by the polyp, and the marginal depression at the base of the polyp may also be accentuated. Superimposed gas in the bowel is excluded by the use of multiple projections. Radiographs must be obtained with the patient in the erect or lateral decubitus position to exclude or confirm the presence of gallstones.

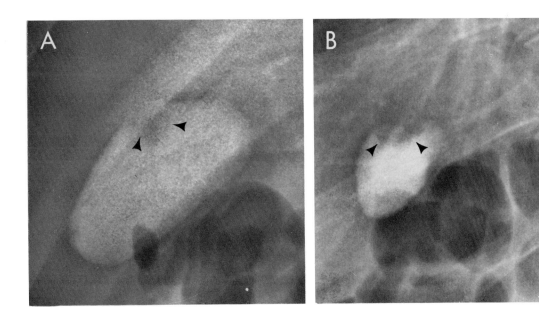

Figure 119.—Cholesterol polyp of the gallbladder.

A 67-year-old woman had chronic dyspepsia.

A, oral cholecystogram, prone oblique projection: Note the lobulated filling defect within the midlumen of the opacified gallbladder.

B, oral cholecystogram, lateral decubitus projection: The intraluminal filling defect in the gallbladder maintains a constant position.

Comment: Cholesterol esters are deposited in the submucosa of this polyp as well as at other sites in the gallbladder. A long, fine stalk attaches the polyp to the gallbladder wall. This is characteristic of cholesterol polyps. It has been proposed that these stalks can become detached and form a nidus for development of a large calculus.

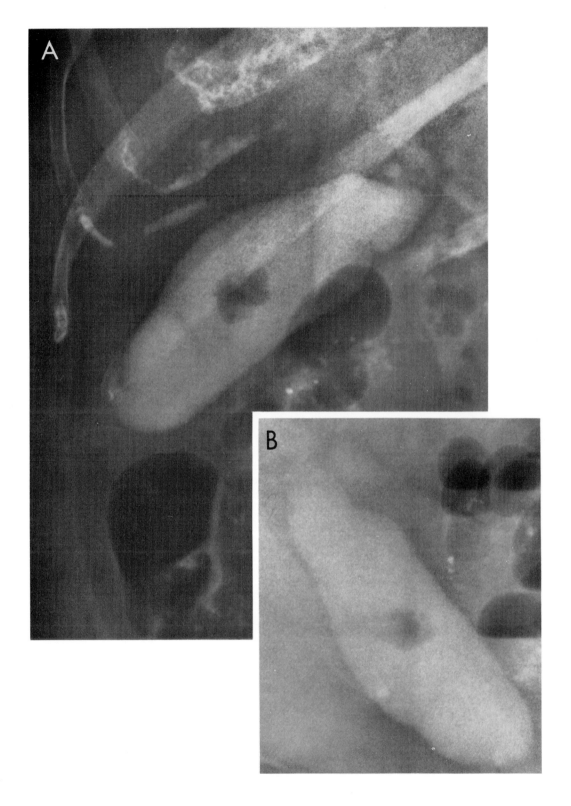

Figure 119 · Cholesterol Polyp / 297

Figure 120.—Polyp of the gallbladder.

A 52-year-old woman had a three-year history of epigastric pain and weight loss.

A, oral cholecystogram, prone oblique projection: There is excellent opacification of the gallbladder and the cystic duct. The common bile duct is partially filled. A radiolucent filling defect is present at the lateral margin of the neck of the gallbladder (**arrow**).

B, oral cholecystogram, prone oblique projection after a fatty meal: The gallbladder has contracted, and the filling defect at the lateral margin near the neck of the gallbladder is better defined than in **A**.

Comment: It was found that this patient's symptoms were caused by a duodenal ulcer. After conservative treatment, the ulcer healed and the symptoms abated. The gallbladder polyp was thus asymptomatic and was not surgically removed. There is much debate as to whether cholecystectomy is advisable whenever gallbladder polyps are diagnosed, symptomatic or not. Carcinoma in situ has been found in a small number of adenomas and papillomas of the gallbladder.

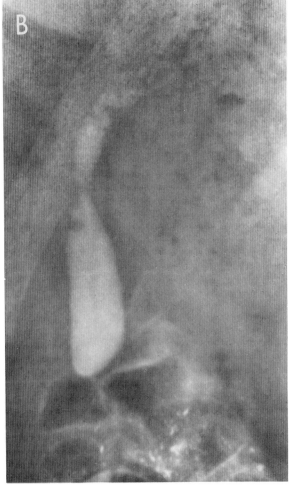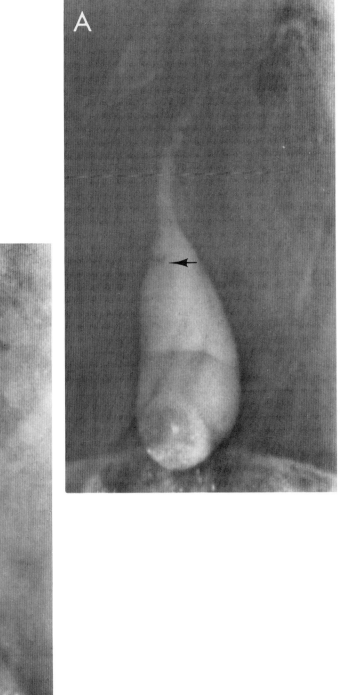

Figure 120 · Polyp / 299

Figure 121.—Polyp of the gallbladder.

A, oral cholecystogram, multiple spot films, anteroposterior erect position with varying degrees of obliquity: The small lobulated filling defects within the lumen of the gallbladder maintain their position in the gallbladder in numerous projections and do not fall to the dependent portion of the gallbladder.

B, oral cholecystogram, prone oblique projection after a fatty meal: The gallbladder has contracted and the polyp is again seen in the same position within the gallbladder (**arrow**). The bile ducts are partially demonstrated after contraction of the gallbladder.

Comment: Surgical proof is not available in this patient, but the radiologic characteristics, such as fixed filling defects, are quite typical for a polyp of the gallbladder.

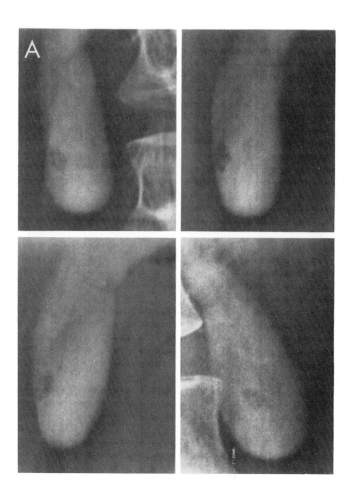

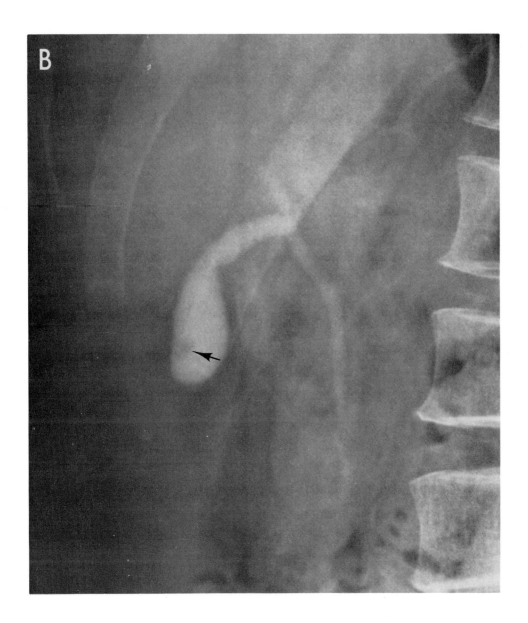

PART 5

The Salivary Glands

Characteristics of Tumors of the Salivary Glands

Tumors of the salivary glands are rare. The reported prevalence of intrinsic salivary gland tumors varies from 0.5–3%. Extrinsic salivary gland tumors that involve the salivary glands in their growth are more frequent. These extrinsic tumors include cervical lymphadenopathy due to tumor metastasis, lymphoma of the cervical lymph nodes, tuberculous or syphilitic lymphadenitis, neurofibroma, lymphangioma, adamantinoma and abscess. The parotid glands are most commonly affected by tumor. The relative incidence is 80% parotid, 15% submaxillary, and 5% sublingual and other minor glands.

Salivary gland tumors usually appear in late adult life with increasing incidence in advanced age. Both sexes are equally affected.

Mixed tumors are the most commonly occurring salivary tumors, constituting 40% of all salivary gland tumors, with approximately 10% being malignant. These tumors are round, nodular, firm masses and are well encapsulated and imbedded in the gland. They are so named because of their epithelial and connective tissue components; the epithelium is arranged in glandular structure. The stroma may resemble cartilage and rarely has calcification within it. Benign mixed tumors recur in 20–30% of instances. This may be a reflection of inadequate surgery rather than a propensity for a malignant change in mixed tumors.

Mucoepidermoid carcinomas are of ductal origin and constitute between 5 and 10% of salivary gland tumors. They are made up of epidermoid cells and mucus-secreting cells. Two distinct grades of mucoepidermoid carcinomas are known: the low-grade carcinoma which is slow growing and runs a relatively benign clinical course, and the high-grade carcinoma which is extremely malignant and fast growing. Regional lymph node metastases are the rule. These tumors are usually poorly encapsulated.

Warthin's tumor is also known as papillary cystadenoma lymphomatosum. It is benign and constitutes 7% of salivary gland tumors. It has been suggested that this tumor arises from branchial cleft rests. A mass of lymphoid tissue grows within the salivary gland, forming spaces within the mass of lymphoid tissue that are lined by eosinophilic epithelium, and the epithelium forms a papillary projection within these spaces. The tumors are completely benign; recurrence or metastasis does not occur. Bilateral Warthin's tumor is occasionally found in the parotid glands.

Adenocarcinoma and squamous cell carcinoma and other rarer types of carcinoma comprise 20–30% of salivary gland tumors. The squamous cell carcinoma is the most rapidly growing. These carcinomas metastasize locally and systemically, and 21% of malignant salivary gland tumors will have metastasized to the cervical nodes at the time of the patient's initial visit.

CLINICAL FEATURES.—Painless swelling is the most common clinical manifestation of salivary gland tumors. With benign tumors, the gradual painless growth is noted over a long period of time. A change in the growth pattern, such as a sudden increase in size, should arouse suspicion of malignancy.

RADIOGRAPHIC FEATURES.—Salivary gland swelling may be noted on routine neck and skull radiographs, utilizing the soft tissue technique. Soft tissue calcification may be noted, but it rarely occurs in mixed salivary gland tumors. Erosion of the mandible is found with 6% of these tumors.

The salivary glands can directly be evaluated by sialography. This technique involves the injection of contrast material, either aqueous or oily, into the duct of a salivary gland after cannulation or catheterization of the duct. With optimal technique, good filling of the ductal system and minimal acinar filling are accomplished.

In extrinsic tumors that involve the salivary glands, the sialogram demarcates the margin of the normal gland from that of the tumor. The general architecture of the gland is preserved. Slight deviation of the intraglandular ducts may occur, but no localized narrowing or kinking of the ducts is seen. The filling defect in the gland is marginal. The neighboring acinar pattern is feathery, but some crowding and compression are evident. If, after stimulation of salivation, a postevacuation radiograph is obtained, complete emptying of the gland will have occurred; no ductal obstruction occurs with extrinsic lesions.

Benign intrinsic salivary tumors are usually well circumscribed and encapsulated. The malignant tumors are irregular and invasive. However, it is not valid to conclude that a well-circumscribed intrasalivary gland tumor must be benign. On sialographic examination the benign tumors cause a filling defect that is well defined, and the ductal system is displaced around the tumor. The ducts are stretched and thinned as they are displaced by the tumor. The tumor outline may be nodular. Focal narrowing and partial obstruction of secondary and tertiary duct branches are seen. On a postevacuation radiograph, segmental intraglandular duct retention may be demonstrated as a result of compression and partial obstruction of a duct.

With the malignant intrinsic tumors, sialography usually shows disruption of the salivary gland architecture, with destruction of acini, bizarre arrangement of the ducts and puddling of contrast material in destroyed areas of the gland. On the postevacuation radiograph, retention of contrast medium in the destroyed area is noted.

SUGGESTED READING

Meine, F. J., and Woloshin, H. J.: Radiologic diagnosis of salivary gland tumors, Radiol. Clin. North America 8:475–485, December, 1970.

Peery, T. M., and Miller, F. N., Jr.: *Pathology: A Dynamic Introduction to Medicine and Surgery* (2nd ed.; Boston: Little, Brown and Company, 1971).

Rubin, P., and Holt, J. F.: Secretory sialography in diseases of the major salivary glands, Am. J. Roentgenol. 77:575–598, April, 1957.

Figure 122.—Normal sialogram.

A, parotid sialogram, puffed-cheek anteroposterior projection: Normal preliminary exposure.

B, parotid sialogram, puffed-cheek, anteroposterior projection: Static injection of Lipiodol using the tabletop screen technique.

C, parotid sialogram, verticosubmental projection: The deep lobe extends medial to the angle of the mandible.

Figure 122, courtesy of Dr. Alexander S. Macmillan, Jr., Massachusetts Eye and Ear Infirmary, Boston.

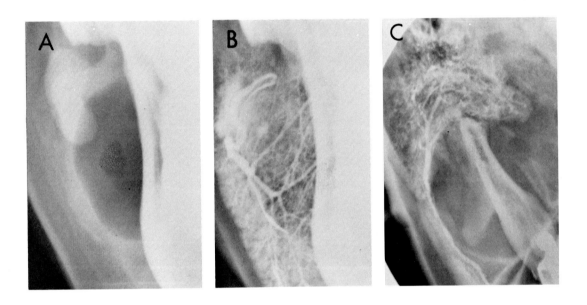

Figure 123.—Normal sialogram.

A, parotid sialogram, submentovertex projection: Stensen's duct runs posteriorly from the opening of the second upper molar tooth in the oral cavity and follows a curvilinear course lateral to the masseter muscle. The parotid gland is variable in size and is divided into superficial and deep lobes by its relation to the ramus of the mandible. There is good opacification of the ductal system, and some acinar filling, as is seen here, is desirable.

B, parotid sialogram, oblique projection: The course of Stensen's duct to the angle of the mandible is practically straight. At this point it curves medially to drain the deep lobe.

C, parotid sialogram, tomogram in lateral projection: The acinar structure is well demonstrated by this technique. The main duct appears to be irregular, but this is due to poor filling at this stage rather than to pathologic change.

Figure 123, courtesy of Dr. Alexander S. Macmillan, Jr., Massachusetts Eye and Ear Infirmary, Boston.

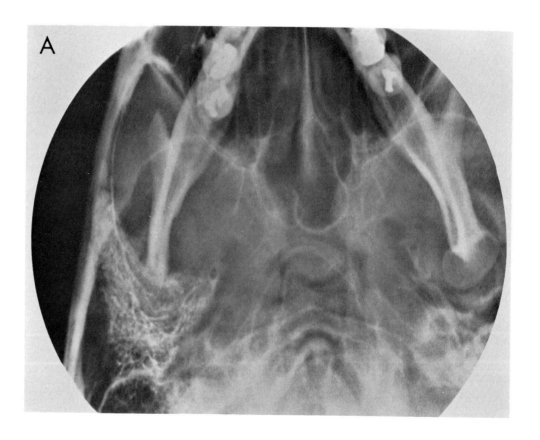

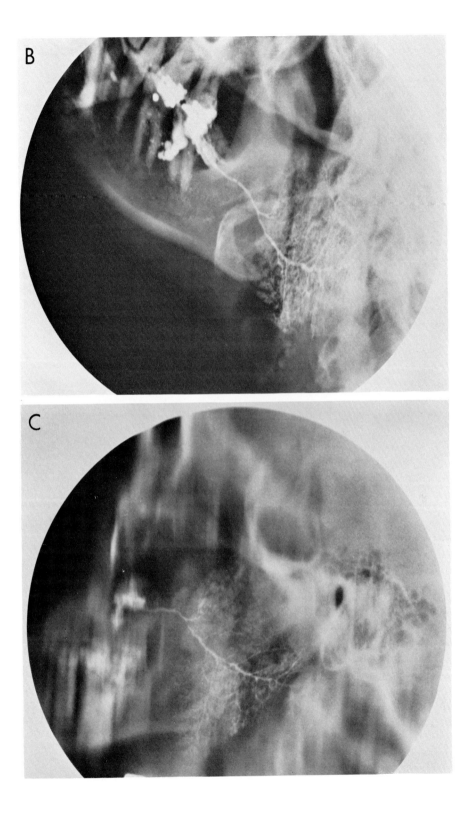

Figure 123 · Normal Sialogram

Figure 124.—Normal sialogram.

A, parotid sialogram: Using the fluoroscopic spot film method, Ethiodol 37% iodine, was injected by way of the Rabinov catheter. Early filling of the ductal system is seen (**arrows**). Some contrast material, as is noted here, may leak into the gauze sponge packing near the opening of Stensen's duct (**left-hand arrows**) and is not to be confused with extravasation into the soft tissues of the cheek.

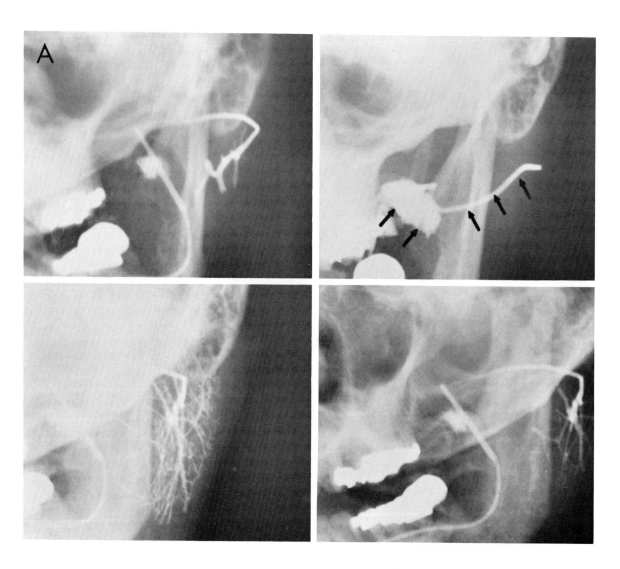

B, right parotid sialogram, anteroposterior projection: In the midfilling phase the ducts are first seen free of bone in the soft tissues of the cheek. The patient is then turned on his left side to provide additional views of the ducts and gland.

(*Continued.*)

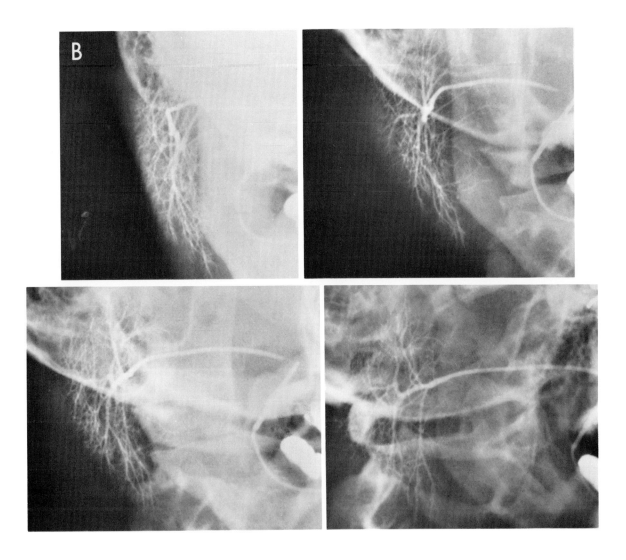

Figure 124 (cont.).—Normal sialogram.

C, right parotid sialogram, early parenchymal phase (**arrows**): The injection is continued until a diffuse parenchymal blush is obtained. In this way both the ducts and the acinar pattern of the gland are used to outline masses.

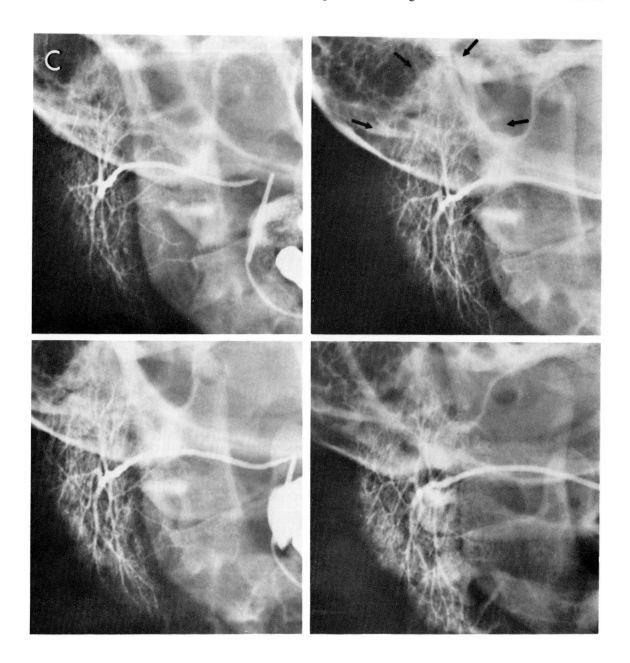

D, right parotid sialogram, late phase **(arrows)**: The injection is continued to the tolerance of the patient and final projections are obtained.

Comment: After this series has been taken, it may be elected to obtain another series of standard plain tabletop projections with the catheter still in place. After removal of the catheter a spot film or dental film of the duct orifice is advisable. Drainage or postevacuation exposures may give additional information.

Figure 124, courtesy of Dr. Alexander S. Macmillan, Jr., Massachusetts Eye and Ear Infirmary, Boston.

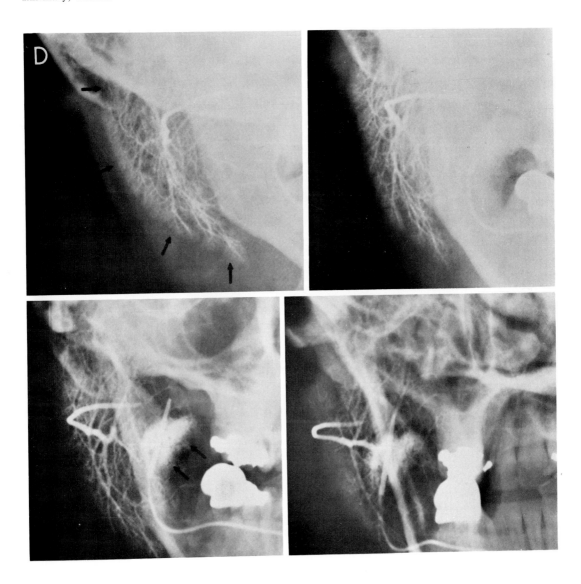

Figure 125.—Normal sialogram.

A, submaxillary sialogram, true lateral projection: Using the tabletop screen technique with the static injection of Lipiodol, the Rabinov catheter is still in place.

B, submaxillary sialogram, oblique lateral projection: This view complements the true lateral projection. Note periapical resorption around the remaining mandibular molar tooth.

C, submaxillary sialogram, verticosubmental projection.

Figure 125, courtesy of Dr. Alexander S. Macmillan, Jr., Massachusetts Eye and Ear Infirmary, Boston.

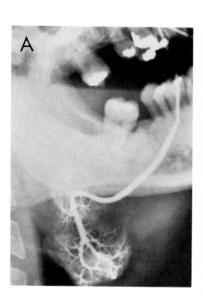
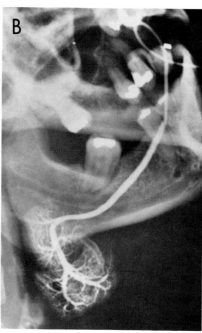
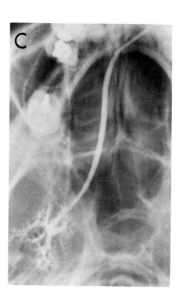

Figure 126.—Lipoma.

Submaxillary sialogram, lateral projection: Note displacement of the ducts (**arrowheads**) around a mass which indents the anterior portion of the gland but shows no evidence of destruction of the glandular architecture. The filling defect is marginal. This appearance is typical of an extrinsic tumor indenting the submaxillary gland.

Figure 126, courtesy of R. Ollerenshaw, BM, DMRD, FRPS, and S. Rose, MD, FRCS, The Royal Infirmary, Manchester, England, and Dental Radiog. & Photog. 29:46, 1956.

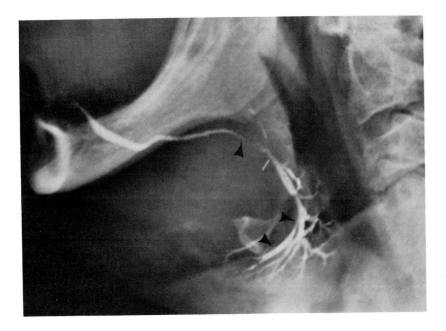

Figure 127.—Neurofibroma within the parotid gland.

A 15-year-old boy was found to have a multinodular mass in the right parotid area during examination for epistaxis and sinusitis. He had a café-au-lait pigmentation in the left antecubital region and several subcutaneous nodules. Biopsy of one of these nodules showed typical findings of von Recklinghausen's disease.

A, parotid sialogram, lateral projection: A multilobulated filling defect occupies the major portion of the parotid gland. Some opacification of acini is seen in the periphery of the gland surrounding the tumor (**arrowheads**). The ducts appear to be displaced and are draped around what appears to be a lobulated benign tumor.

B, parotid sialogram, modified anteroposterior projection: Note stretching and thinning of the ducts as they course around the tumor. Some normal acini are demonstrated in the surrounding gland.

Figure 127, courtesy of Dr. Alexander S. Macmillan, Jr., Massachusetts Eye and Ear Infirmary, Boston.

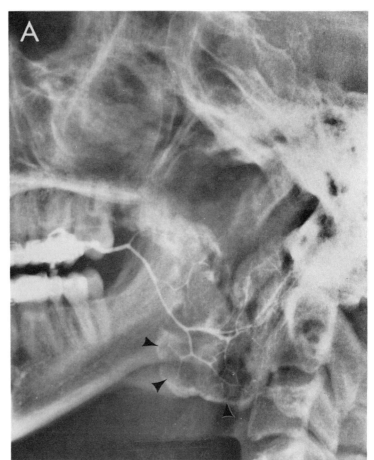
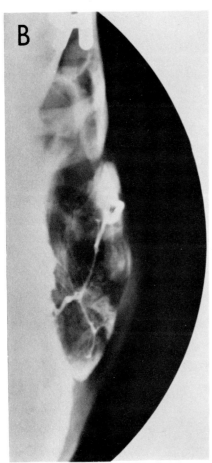

Figure 127 · Parotid Neurofibroma / 317

Figure 128.—Mixed salivary tumor.

A 14-year-old girl had a small movable mass behind the angle of the mandible that had been present for six to eight weeks.

A, left parotid sialogram, submentovertex projection: A well-marginated filling defect involving both the superficial and the deep lobes of the parotid gland is seen (**arrows**). There is some stretching of the secondary ducts in the region of the tumor, but no kinking or obstruction is seen.

B, left parotid sialogram, tomogram in lateral projection: The main duct is depressed inferiorly by the mass which lies in the superior portion of the gland (**arrows**). The normal acini in the surrounding gland are well demonstrated.

C, left parotid sialogram, modified anteroposterior projection: Note ductal displacement.

Figure 128, courtesy of Dr. Alexander S. Macmillan, Jr., Massachusetts Eye and Ear Infirmary, Boston.

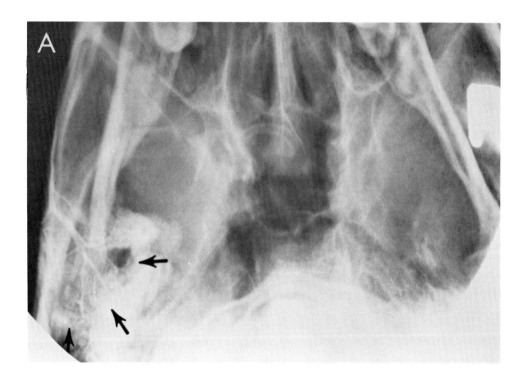

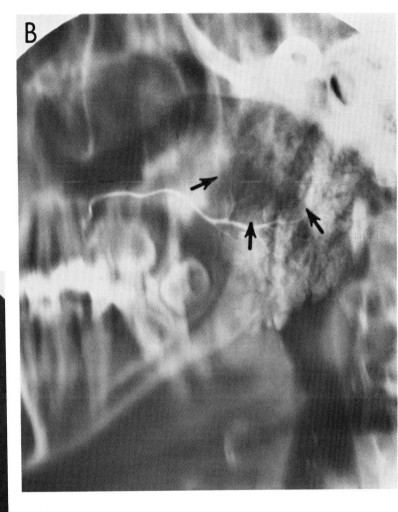
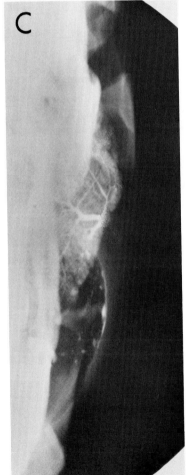

Figure 128 · Mixed Salivary Tumor / 319

Figure 129.—Mixed salivary tumor.

Parotid sialogram, lateral projection: Note the small circumscribed filling defect in the lower portion of the parotid gland (**arrows**). There is some displacement and draping of the ducts around this mass, but there is no kinking or obstruction of the ducts. The acinar pattern of the parotid gland contiguous to the tumor is normal.

Figure 129, courtesy of Dr. Keith Rabinov, Cardinal Cushing Hospital, Brockton, Mass.

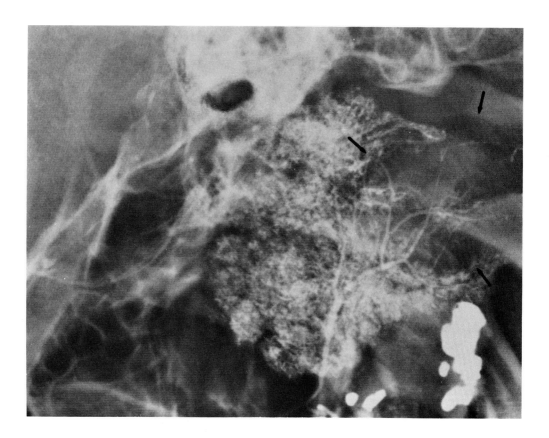

Figure 130.—Mixed salivary tumor.

Parotid sialogram, oblique projection: The secondary and tertiary duct branches are displaced by the smooth, well-marginated filling defect in the deep portion of the gland. The gland contiguous to the filling defect is normal in appearance. The smooth margins (**arrowheads**) are defined by the normal acinar pattern in the contiguous gland.

Figure 130, courtesy of Dr. Keith Rabinov, Cardinal Cushing Hospital, Brockton, Mass.

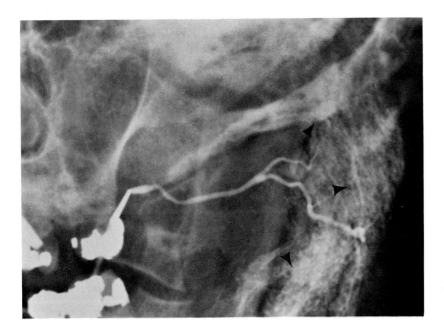

Figure 131.—Mixed salivary tumor.

A 53-year-old woman had a one-month history of swelling in the region of the right parotid gland. The mass measured approximately 3.5 × 2.5 cm. The facial nerve was intact.

A, right parotid sialogram, oblique projection: Note the filling defect in the posterior portion of the gland. This is well circumscribed and smoothly outlined (**arrowheads**). The ducts are displaced in a curvilinear course by the tumor. No evidence of invasion of the contiguous portion of the gland is seen.

B, right parotid sialogram, submentovertex projection: On this projection the mass is found to involve both the superficial and deep lobes of the parotid gland (**arrows**).

Figure 131, courtesy of Dr. Alexander S. Macmillan, Jr., Massachusetts Eye and Ear Infirmary, Boston.

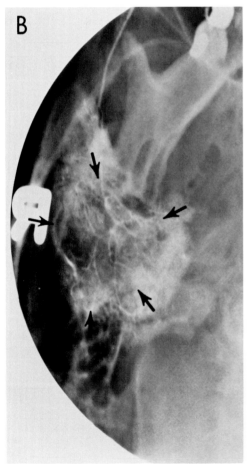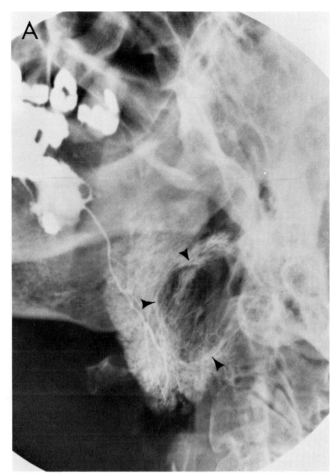

Figure 131 · Mixed Salivary Tumor

Figure 132.—Mixed salivary tumor.

A 75-year-old woman had a firm, nontender mass in the area of the right parotid gland.

Parotid sialogram, oblique projection: The main duct runs a curvilinear course around a mass within the salivary gland with stretching of the secondary duct branches. Little acinar filling has been accomplished because of the large size of the tumor. No kinking or obstruction of the ducts has occurred.

Figure 132, courtesy of Dr. Alexander S. Macmillan, Jr., Massachusetts Eye and Ear Infirmary, Boston.

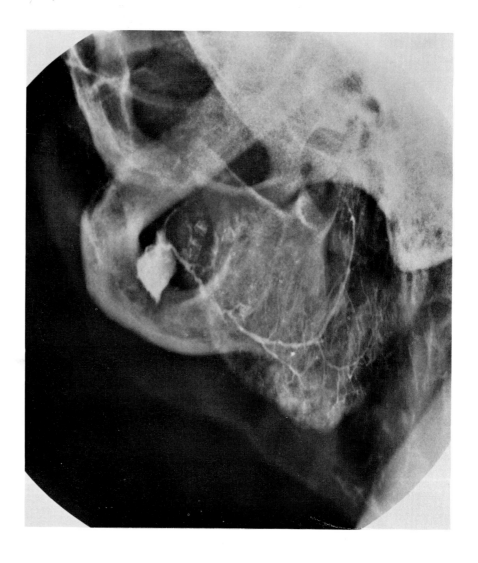

Figure 133.—Mixed tumor of parotid gland.

A, parotid sialogram, anteroposterior projection: Using the static Lipiodol technique, ducts are seen draped around the tumor (**asterisk**).

B, parotid sialogram, lateral projection: The tumor is again identified (**asterisk**).

C, parotid sialogram, verticosubmental projection: The mass (**asterisk**) is in deep and superficial lobes.

Figure 133, courtesy of Dr. Alexander S. Macmillan, Jr., Massachusetts Eye and Ear Infirmary, Boston.

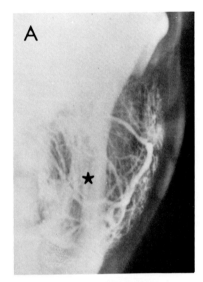
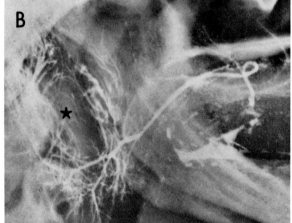
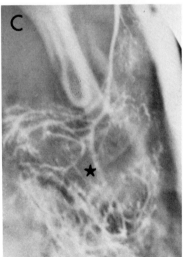

Figure 134.—Malignant invasion of the parotid gland by a pharyngeal lymphosarcoma.

Parotid sialogram, anteroposterior projection: Note marked distortion of the ductal branches, including stretching, segmental narrowing and dilatation with complete obstruction. A large mass occupies a major portion of the gland. A small number of contrast medium puddles are noted within the gland, and acinar filling is also present. These findings indicate invasion and destruction of the gland. This radiographic appearance is typical of a malignant lesion.

Figure 134, courtesy of R. Ollerenshaw, BM, DMRD, FRPS, and S. Rose, MD, FRCS, The Royal Infirmary, Manchester, England, and Dental Radiog. & Photog. 29:46, 1956.

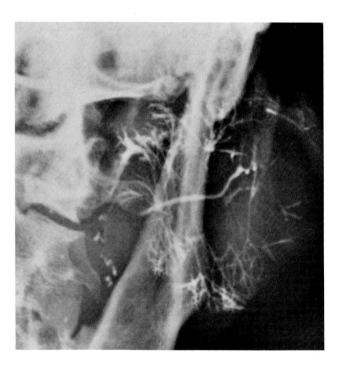

Figure 135.—Tuberculous intraparotid lymph glands.

Parotid sialogram, anteroposterior projection: A bilobular filling defect is present in the inferior and medial portion of the superficial lobe of the parotid gland (**arrowheads**). This is a well-defined lesion with no evidence of ductal abnormality apart from displacement. The general architecture of the gland is well preserved, and the neighboring portions of the parotid gland are normal.

Figure 135, courtesy of R. Ollerenshaw, BM, DMRD, FRPS, and S. Rose, MD, FRCS, The Royal Infirmary, Manchester, England, and Dental Radiog. & Photog. 29:45–46, 1956.

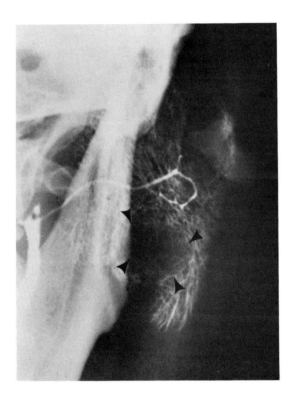

Figure 136.—Inflammatory mass extrinsic to submaxillary gland not involving gland.

Submaxillary sialogram, lateral projection: A lead shot has been placed on the skin surface over the palpable mass. Excellent opacification of the submaxillary gland has been obtained. The submaxillary gland sialogram is normal. Note the soft tissue mass anterior to the submaxillary gland.

Figure 136, courtesy of Dr. Keith Rabinov, Cardinal Cushing Hospital, Brockton, Mass.

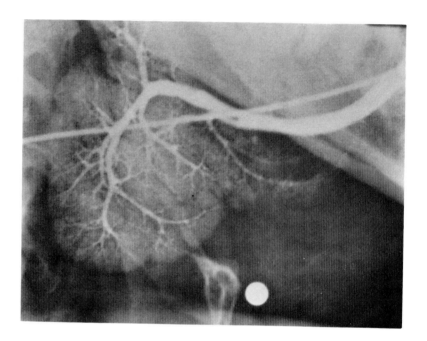

Index

A

ADENOCARCINOMA, *see* organ involved
ADENOMA, *see* site involved
AMPULLA OF VATER
  carcinoma
    characteristics, 229-32
    radiographic features, 231, 254-61
  villous adenoma, 262-63

B

BILE DUCTS, EXTRAHEPATIC
  carcinoma
    characteristics, 229-32
    cholelithiasis with, 246-47
    clinical features, 230-31
    metastases from, 230, 248-49
    radiographic features, 231-32, 234-61
    —angiography, 231
    —cholangiography, 231
    —fiberoptic endoscopy, 231
  common bile
    carcinoma, 234-43, 251
    —characteristics, 229-32
    chronic inflammation resembling tumor, 269
    dilated, on endoscopy, 272-73
    normal, on endoscopy, 270-71
    with pancreatic carcinoma, 4, 60-61, 74-77
    —percutaneous transhepatic cholangiography for, 4-5, 57
  hepatic
    carcinoma, 229-32, 244-53
    compression of common duct due to metastases, 268
  papilloma, 232
    and biliary carcinoma, 229
  tumor characteristics, 229-32
    benign, 232
    primary malignant, 229-32
BILIARY TREE
  see also Bile Ducts
  papilloma of, and biliary carcinoma, 229
BREAST: liver metastasis from carcinoma, 158-59, 162

C

CARCINOID: metastasis to liver, 161
CARCINOMA, *see* organ involved
CHOLANGIOCARCINOMA, HEPATIC, 123, 156-57
CHOLECYSTITIS GLANDULARIS PROLIFERANS, 280
CHOLEDOCHOCELE, 232, 264-67
CHOLELITHIASIS
  with biliary duct carcinoma, 246-47
  and gallbladder carcinoma, 277
  and gallbladder papilloma, 279
COLON: liver metastases from carcinoma, 163-76
CYST(s)
  choledochal, 232, 264-67
  hepatic, 127-29, 200-21

D

DIABETES MELLITUS: and pancreatic carcinoma, 3
DUODENUM
  with ampullary carcinoma, 255
  with gallbladder carcinoma, 286-89
  with pancreatic carcinoma, 4, 56, 58-60, 64-65, 67, 68, 69, 70-71, 74-75

E

ESOPHAGUS: with pancreatic carcinoma, 66

F

FIBROSARCOMA: of thigh—metastases to liver, 178-79

G

GALLBLADDER
  adenoma, 279, 295
  adenomyoma, 280
  carcinoma
    calcification and, 278
    cholelithiasis and, 277
    clinical features, 277-78

GALLBLADDER (cont.)
    metastases from, 277, 282-91
    pathology, 277
    in polyps, 292, 298
    radiographic features, 278, 282-94
  metastases to, 279
  with pancreatic carcinoma, 4, 57, 74-77
  papilloma, 279
  polyps
    asymptomatic—cholecystectomy
      for, 292
    carcinoma in, 292, 298
    characteristics, 279-80
    cholesterol polyps, 279, 296
    clinical features, 280
    inflammatory, 280
    radiographic features, 280, 295-301
  sarcoma, 278-79
  tumor characteristics, 277-80

H

HAMARTOMA: of liver, 127, 188-89
HEMANGIOMA: hepatic, 125-26
HEPATOMA, 121-23, 132-55

I

INSULINOMA, 7-8, 90-91, 96-97
ISLETS OF LANGERHANS, 7
  tumors, 7-8, 86-97

J

JAUNDICE
  with biliary duct carcinoma, 230
  with gallbladder carcinoma, 277, 290-91
JEJUNUM: liver metastasis from
    leiomyosarcoma of, 177

K

KIDNEY
  carcinoma—metastases to pancreas,
    114-17
  polycystic disease with polycystic
    disease of liver, 128, 220-21

L

LEIOMYOSARCOMAS of jejunum—liver
    metastasis from, 177
LIVER
  abscess
    characteristics, 129-30
    radiographic features, 130, 192-99
  adenoma (benign), 126-27

    liver cell type, 190-91
  calcification, in metastases, 125
  cholangiocarcinoma
    angiography, 123
    characteristics, 123
    cholangiography, 123
    radiologic features, 123, 156-57
  cysts, 127-29
    hydatid, 128-29, 212-17
    polycystic disease, 128, 218-21
    simple, 127-28, 200-11
  granulomas—diffuse chronic, 222-23
  hamartoma, 127
    with cystic degeneration, 188-89
  hemangioendothelioma, 126
  hemangioma(s)
    angiography, 126
    cavernous, 125-26, 180-87
    characteristics, 125-26
    radiologic features, 126, 180-87
  hepatoma
    angiography, 122-23
    characteristics, 121-23
    in infancy, 122
    pharmacoangiography, 122
    radiographic features, 122, 132-55
  metastases to, 124-25, 158-79
    from biliary carcinoma, 230, 248-49
    from breast, 158-59, 162
    from carcinoid, 161
    from colon, 163-76
    from fibrosarcoma of leg, 178-79
    from gallbladder, 277, 290-91
    from jejunal leiomyosarcoma, 177
    from pancreatic adenocarcinoma,
      161
    from stomach involving hilar nodes
      and compressing common hepatic
      duct, 268
    vascular categories, 124
  regenerated nodule, 224-25
  tumors
    benign, 125-130
    characteristics, 121-30
    metastatic, 124-25
    —angiography, 124-25
    —calcification in, 125
    —radiologic features, 124-25, 158-79
    —radionuclide studies, 125
    —vascular categories, 124
    primary malignant, 121-23
LYMPHOSARCOMA: pharyngeal, invading
    parotid, 326

M

MELANOMA: metastatic to gallbladder,
    279

P

PANCREAS
   adenocarcinoma
      angiography 5-6
      —direct injection, 26
      —Priscoline to enhance venous filling, 22
      characteristics, 3-6
      cholangiography (percutaneous transhepatic) in, 4-5, 57
      clinical features, 4
      diabetes mellitus and, 3
      metastases, 3
      —to liver, 160
      and pancreatitis, 3, 6
      pathology, 3
      radiographic features, 4-6, 12-76
      radionuclide studies, 5, 77-79
      selenomethionine scans, 5, 77-79
   calcification
      with cysts, 9, 106-109
      with neoplasm, 60-61, 84
   cyst(s) 8-9
      radiographic findings, 9, 98-113
   cystadenocarcinoma, 6-7
   cystadenoma, 6-7, 80-85
   islet cell tumors, 7-8, 86-97
      adenoma, 7
      carcinoma, 88-89, 92-95
      insulinoma, 7-8, 90-91, 96-97
      —Whipple's triad with, 7
      radiographic findings, 8
      Zollinger-Ellison syndrome with, 7
   metastases from, 3, 160
   metastasis to, from renal cell carcinoma, 114-17
   normal—selenomethionine scan, 77
   pseudocyst(s), 8-9
      pancreatitis and, 8
      radiographic findings, 9, 98-105, 108-11
   tumor characteristics, 3-9
PANCREATITIS
   angiographic picture, 6
   differentiation from atherosclerosis, 6
   with pancreatic carcinoma and diabetes, 3
   and pseudocysts, 8
PAPILLARY CYSTADENOMA LYMPHOMATOSUM, 304

R

RADIONUCLIDE STUDIES
   of liver metastases, 125, 164-65, 168-69, 176
   of pancreas
      adenocarcinoma, 5, 77-79
      normal, 77

S

SALIVARY GLANDS
   carcinoma
      mucoepidermoid, 304
      squamous cell, 305
   inflammatory mass extrinsic to submaxillary gland, 328
   lipoma, 315
   metastases to, 304, 326
   neurofibroma, 304, 316-17
   normal sialograms, 307-14
   parotid
      metastases to, 304, 326
      mixed tumors in, 318-25
      neurofibroma, 316-17
      tuberculous intraparotid lymph glands, 327
   submaxillary—extrinsic inflammatory mass, 327
   tuberculous intraparotid lymph glands, 327
   tumors
      characteristics, 304-06
      clinical features, 305
      mixed, 304, 318-25
      radiographic features, 305
      sialography, 305, 315-26
      Warthin's, 304
SELENOMETHIONINE SCANS (PANCREATIC)
   for carcinoma, 5, 77-79
   normal pancreas, 77
SIALOGRAPHY
   of extrinsic submaxillary inflammation, 327
   of normal salivary glands, 307-14
   of salivary gland tumors, 315-36
   technique, 305
   of tuberculous intraparotid lymph glands, 327
STOMACH
   carcinoma—metastasis to hepatic hilar nodes, compressing common hepatic duct, 268
   with pancreatic carcinoma, 4, 62-66, 69, 72

W

WARTHIN'S TUMOR, 304
WERMER'S SYNDROME: with islet cell tumors, 7

Z

ZOLLINGER-ELLISON SYNDROME: with islet cell tumors, 7-8